Handbook of Energy Efficiency for Industry

The Author

Dr Ramesh Bhatia is an engineer (Electrical and Mechanical) from Naval College of Engineering and MBA from Jamnalal Bajaj Institute, Mumbai. With total experience of over 40 years in reputed organizations, he has worked in manufacturing, projects, maintenance and business management. He is an expert for measurement and verification of energy consumption. He has over 15 years of hands on experience in the practice of energy efficiency and conservation. For his work he has covered almost all the major and medium industries including power plants, cement , textile and other manufacturing concerns. In this endeavour he has interacted with senior management and addressed the challenges faced in reducing energy consumption in their plants.

Handbook of Energy Efficiency for Industry

Ramesh Bhatia

2019
Associated Publishing Company®
A Division of
Astral International Pvt. Ltd.
New Delhi – 110 002

ISBN: 9789389569193 (Int. Edition)

Publisher's Note:

Every possible effort has been made to ensure that the information contained in this book is accurate at the time of going to press, and the publisher and author cannot accept responsibility for any errors or omissions, however caused. No responsibility for loss or damage occasioned to any person acting, or refraining from action, as a result of the material in this publication can be accepted by the editor, the publisher or the author. The Publisher is not associated with any product or vendor mentioned in the book. The contents of this work are intended to further general scientific research, understanding and discussion only. Readers should consult with a specialist where appropriate.

Every effort has been made to trace the owners of copyright material used in this book, if any. The author and the publisher will be grateful for any omission brought to their notice for acknowledgement in the future editions of the book.

Published by : **Associated Publishing Company®**
A Division of
Astral International Pvt. Ltd.
– ISO 9001:2015 Certified Company –
4736/23, Ansari Road, Darya Ganj
New Delhi-110 002
Ph. 011-43549197, 23278134
E-mail: info@astralint.com
Website: www.astralint.com

Digitally Printed at : **Replika Press Pvt. Ltd.**

Web www.teriin.org

The Energy and Resources Institute
Darbari Seth Block
I H C Complex
Lodhi Road
New Delhi – 110 003
Tel. 2468 2151 or 2468 2152
E-mail dg@teri.res.in
Fax 2468 2144
India +91 • Delhi (0) 11

Ajay Mathur *Ph.D.*
Director-General

Foreword

Industry is the single largest energy consuming sector in India, and detailed energy audits (across all sizes of industries) indicate a huge variation in energy usage even among industries that produce the same or similar products. This diversity in energy usage is both troubling, as well as challenging. On the one hand, it reflects the simultaneous presence of various vintages of technologies, with older vintages exhibiting lower energy efficiency; however these older, less efficient units are able to compete in the market because of their lower, and sometimes completely depreciated capital costs. On the other hand, there are also large variations in energy usage amongst industrial units manufacturing similar products and of a similar technological vintage. This second kind of diversity, most often seen in SMEs, suggests that industrial units with lower energy efficiency either do not know the potential for energy and costs savings, or are unwilling to be "guinea pigs" in trying out newer, but more energy efficient technological options.

This "Handbook for Energy Efficiency in Industries" addresses available opportunities and their viability directly, covering important topics that are relevant for enhancing energy efficiency in utilities in industries - the presence of this Handbook on the bookshelves of industry managers will, at least, indicate that they are aware of both the opportunities that exist, and of methodologies to assess the viability of those opportunities. Further, the approaches, applications, practices, and benefits that are provided in this handbook are based on actual operating experience and gleaned through the many years of Cdr. Bhatia's experience across the entire breadth of utilities in the energy efficiency industry. They represent, therefore, the state-of-the-art in terms of what is achievable; the term "Handbook" for this monograph is indeed appropriate.

I had the pleasure to working very closely with Cdr. Bhatia when he was with Conzerv and Schneider Electric, and I was with the Bureau of Energy Efficiency in the Government of India. During our interactions, I gained from his practical knowledge and experience, and I am sure that all readers of this Handbook would similarly benefit from this product of his experience and expertise.

(Ajay Mathur)

Regional centres within India

North-Eastern Regional Centre
Guwahati
Tel. (361) 233 4790
E-mail terine@teri.res.in
Fax (361) 233 4869

Southern Regional Centre
Bangalore
Tel. (80) 2535 6590 (5 lines)
E-mail terisrc@teri.res.in
Fax (80) 2535 6589

Goa
Tel. (832) 245 9306 or 245 9328
E-mail terisrc@teri.res.in
Fax (832) 245 9338

Western Regional Centre
Mumbai
Tel. (22) 2758 0021 or 4024 1615
E-mail terimumbai@teri.res.in
Fax (22) 2758 0022

Himalayan Centre
Mukteshwar
Tel. (5942) 286 433
E-mail praveen.sharma@teri.res.in
Fax (5942) 286 460/433

TERI worldwide

TERI **North America**
Tel. (+91) 9811150290
E-mail annapurna.vancheswaran@terina.org

TERI **Europe**
London
Tel. (+44 20) 8947 9145
E-mail rkumarrobins@gmail.com

Antwerp
TERI-VITO
Tel. (+32 3) 286 74 31
E-mail arun.jacob@grihaindia.org

Utrecht
Utrecht University, The Netherlands
Tel. (+31 030) 253-5801
E-mail f.romein@uu.nl

Nordic
University of Eastern Finland
Tel. (+358) 40 7470711
E-mail pradipta.halder@uef.fi

TERI **Japan**
Tokyo
Tel. (+81 3) 3519 8970
E-mail teri@iges.or.jp
Fax +81 33 5195 1084

TERI **Africa**
Ethiopia
Tel. +251 (0) 936 56 6641
E-mail rita.effah@teri.res.in

TERI **Gulf**
Dubai, UAE
Tel. (+971 50) 865 0912
E-mail meenasj@teri.res.in

An ISO 9001:2008 certified organization

Preface

It is estimated that nearly 35% of energy consumption in India is attributable to industries. Government of India through Bureau of Energy Efficiency(BEE) has initiated many schemes for energy efficiency in industries in the last decade. While most of these schemes are voluntary in nature, there is Perform, Achieve and Trade (PAT)scheme which is mandatory for energy intensive industries with a stick and carrot approach. However, resultant effect of these schemes is yet to be evaluated.

The main driver for energy efficiency in industries is the commercial aspect which translates into competitive edge. Cost of energy consumption per unit of production determines the relative competitive position of a company. Cost of energy varies immensely in industries with higher component of energy in cost of production in Aluminum, Caustic Soda, Cement among others. Relatively lower percentage of energy in production cost is in FMCG or textile sector among large industries.

Most of the industries - large, medium or small- have at one time or the other taken the route of energy audit to address energy efficiency. Due to various reasons, only a small proportion of recommendations brought out in energy audit reports are given effect as obstacles come in the way at different levels – technical, managerial or ownership.

This book is an attempt to create awareness for energy efficiency in industries specifically focusing on utilities which are common to all industries. As processes for production vary from industry to industry and even within a sector, most often there is need for technological intervention which can only be obtained externally from the concerned organizations and usually at considerable investment. Utilities on the other hand being common, technical improvements or even replacements can be more easily accessed by the industry and its management at relatively reasonable cost.

The approach in writing the book is to make it most widely read and understood at all levels in the industry from lower technical staff to senior management looking for internal solutions to obtain energy efficiency in their production areas. This has been done by avoiding intricate technical content and laying more emphasis on simplification.

Presuming that general information published by BEE and NPC is for public consumption, diagrams and some illustrations have been sourced from such communications.

Author

Contents

1

Energy Efficiency in Industrial Utilities

Due to steady increase in energy pricing, there is increasing awareness of need for energy efficiency for commercial reasons as the energy price increases result in higher cost of production. While the industry management is able to make some effort for reduction in cost of production due to raw material and other components, reducing energy consumption, presents the industry with a more difficult and technically complex proposition with no easy solutions available. While a spate of energy audits conducted in industry in the last decade enhanced the awareness for energy efficiency, the solutions recommended by energy auditors did not always convince the managements to invest in such measures.

Thus, there is a trust deficit in the industry with suggested solutions which probably developed due to some of the energy auditors going overboard with expected energy saving figures and the industry ownership not being adequately knowledgeable about EE. Quite often, some industry personnel require self help guideline to take their own measures for EE. Management also needs to be aware of possible avenues for energy efficiency within their industry. Due to this reason, there is a focus on energy efficiency in industry with particular emphasis on industrial utilities which are common to most of them.

Management should aim at effective use of energy for commercial reasons so that the profitability is maximized. With energy efficiency in production and utilities in particular, it is possible to reduce Specific Energy Consumption (SEC) per unit of output in a production unit. Benchmarking and comparison within a similar industry helps to work towards the energy efficiency at optimum level.

Common Industrial Utilities

Utilities and systems which are most common in industries are:

Compressed air system

Boilers and steam distribution

Refrigeration and air conditioning

Distribution of electricity within the plant

Power factor, harmonics and other power quality aspects

Electric motors

Pumps and blowers

Transformers

Fuels and combustion

Waste heat recovery

Cooling towers and effluent treatment

Alternate sources of electricity

Renewables and environment

Energy monitoring and utilizing ISO 500001

Industrial buildings

Auxiliary power consumption

Most of the industrial plants have 10 to 12 of the utilities/systems listed above.

ISO 50001

This is a unique International standard developed for energy efficiency in any type of organization. The methodology provided in the standard helps to implement energy efficiency and achieve a self declared objective of reducing energy consumption. Thus adherence to ISO 50001 helps in achieving and improving energy efficiency in an industrial unit.

Energy Audit

This is a common procedure which helps to indicate avenues in any plant, industry, building or organization the scope for energy saving which may be possible in specific area. It is a necessary first step before going further to achieve energy efficiency and reducing consumption.

Energy audits can have a wide variance in quality and the charges levied. Professional auditors with good standing and antecedents should be selected for the assignment. When an energy audit report is received it should be thoroughly analysed for implementation to obtain benefit of the exercise. Quite often the auditors are willing to undertake the implementation and they can be committed to prove the recommendations made in their report.

Some simple dos and don’ts for energy efficiency

- Output from a pump should not be throttled. It is better to use variable speed drive, impeller trimming or as a permanent measure resize the pump.
- Similarly, blower output should not be controlled by using dampers. Better to use variable speed drive, impeller trimming or resizing the blower.
- Eliminate compressed air leakages, reduce cut off pressure and as a long term measure avoid use of compressed air wherever possible by using mechanical or electrical means.
- Steam leakages should be eliminated, output pressure from the boiler reduced to minimum relating it to pressure which is necessary at the driven end.
- Use automatic controls for all machinery and equipment.

In the subsequent chapters energy efficiency in each utility is discussed and focus is on the possible steps which can be locally undertaken in the plant for this purpose.

2

Distribution of Electricity in the Plant

All industrial plants have electrical distribution to run the machinery and other production equipment within the plant. Distribution may be carried out at LT (440 V) or higher voltages like 11 KV. Most of the small and medium size plants with loads of 20 to 30 KVA would operate on LT whereas those with MVA loads such as cement plants etc may use higher voltages. Power plants, steel plants and heavy engineering plants may use multiple voltage levels. The purpose is to keep the electricity distribution losses to minimum. Quite often loads are added subsequent to original distribution plan without adding appropriate distribution equipment or cables. Any major loads should be added after due consideration as it can lead to higher line losses.

Electricity Billing

Billing from electricity supplier may be done based on sanctioned load as well as energy consumption (in units of electricity).If the billing includes load charges separately, sanctioned load or contracted demand should be kept to minimum providing for occasional heavy consumption so that such occurrences do not invite penalties for exceeding maximum demand. Demand can be managed by:

- Containing the demand within the level sanctioned by electricity board by maintaining KVA load by various means.
- There should be staggered or sequential starting of heavy motors so that starting current can be kept within limits.
- Non essential loads can be switched off when the demand is likely to exceed sanctioned level.

Maximum Demand

In a plant, demand will vary continuously with time. For determining maximum demand, average of demand in an interval fixed by supplier is essential. If the average in the given demand in say 30 minutes for example, exceeds contracted demand, supplier may levy a penalty. It should be borne in mind that it is not the peak which matters but the average as shown in the diagram below:

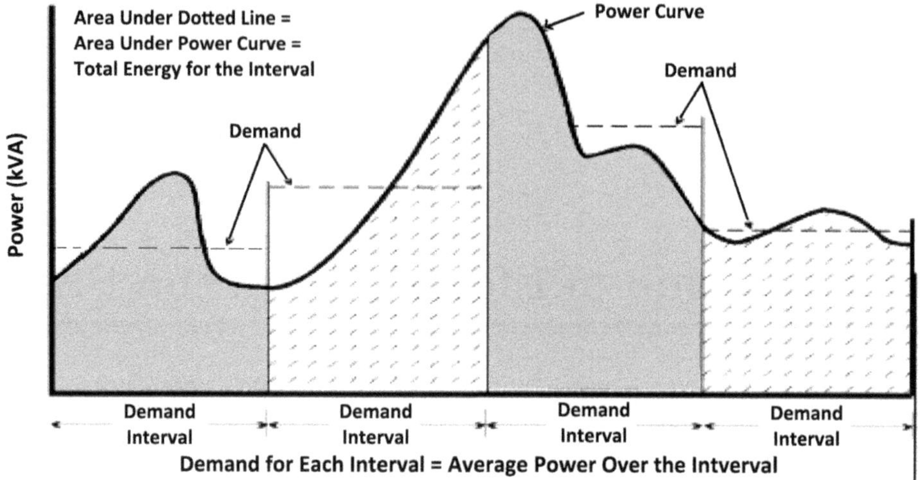

Courtesy BEE

Power Factor

Power factor affects the consumer industry in two ways. Electricity supplier may levy a penalty for low PF. On the other hand, low PF in electrical distribution results in higher line losses as cables carry higher level of current.

To avoid penalty, PF is improved by providing capacitors at the supply point. This arrangement does not improve PF in internal distribution. PF in distribution can be improved by adding small value capacitors at loads which cause the PF to deteriorate. For example, induction motors which are most commonly used in plants, have an operating PF of around 0.8 lagging. In each large induction motor, a capacitor of appropriate value can be added to improve PF. Similar arrangement would be valid for other loads having low PF. Consequently, PF at supply point would be better and require capacitors of lower value there.

Location of capacitors and maintaining PF

- Main distribution boards (Unity)
- Sub distribution boards (0.95)
- Load ends (0.85)

Benefits of improving PF

- Avail rebates from SEB
- Reduced heat loss of network
- Voltage improvement at load end

- ➢ Reduced maximum demand charges

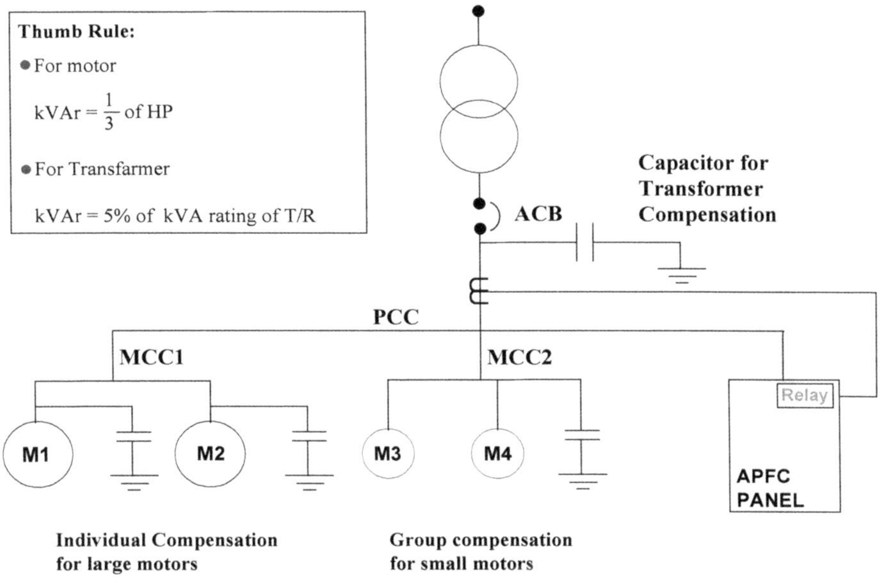

Courtesy BEE

Factors effecting location of capacitors

While the primary aim of PF capacitors is to improve PF, simultaneously the added advantage is of lesser current flowing through the cables and consequent reduction in losses and heat generation. The factors effecting location of capacitors are as follows:

- ➢ Location of capacitors near the mains to achieve unity PF ensures that penalty by electricity supplier is avoided.
- ➢ Apart from penalty avoidance, benefit of locating capacitors at mains, in fact, accrues to the utility only. Load losses within the plant are not affected.
- ➢ To minimise load losses, ideally each load contibuting to low PF (all loads except resistive heating) should have compensating capacitors located there.
- ➢ When the current flowing through the conductors is reduced, the voltage at load end also improves.
- ➢ It should be taken into consideration that the motors are provided capacitors below no load KVAR of the motor. Higher value capacitor can cause damage.

- Location of capacitors with switching arrangement is desirable for power centres or feeders.

Indicators for losses in distribution

- If any cable or conductor is overheated (if outer surface is hot it is obvious that the conductor itself is overheated) then loss in distribution is indicated,
- If the voltage at the load end is low then it shows the loss taking place within the length of the cable.
- Measurement of distribution loss is very important.The most accurate and reliable method is by actual measurement of loss. While the input watts are known from the supply meter (which is also used for billing purpose), the more difficult measurement is that of actual consumption as the difference would provide the level of losses in distribution.
- To overcome this situation and arrive at the figure of accurate loss, the only way out is by simultaneous measurement of input and output watts. This can be done by on line measurement from meters or devices connected to both input and output.
- In conclusion, it can be said that loss measurement in distribution and monitoring is essential. On line measurement can provide the accuracy and convenience.

3

Efficient Running of Motors

On an average, 70 to 80 percent electricity in a plant or industry is consumed by electric motors. As such it is important to focus attention on efficient running of motors. Of the various types of motors, induction motors are the most common. Inefficient operation of motors in industry is usually due to following reasons:

- Under loading or over loading
- Repeated rewinding
- Inefficient motors
- Phase imbalance and improper voltage levels

Effect of using higher than required rating

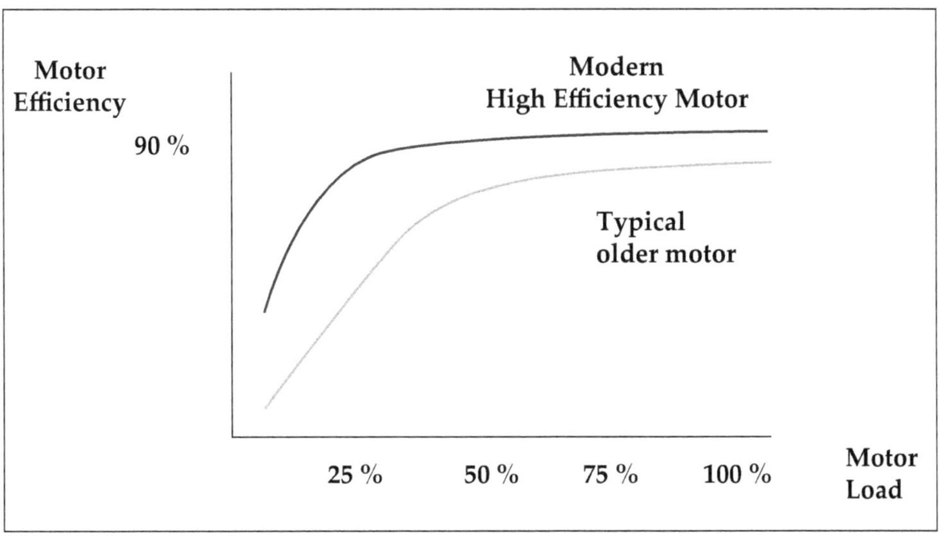

Courtesy BEE

Parameter	Case 1	Case 2	Case 3
Motor Load	7.5 kw at F.L.	10 lw at ¾	15 kw at ½ load
Required load	7.5	7.5	7.5
Motor EFF (%)	88	84	79
Motor input kw	8.5	9.0	9.5
Unit consumed kwh	42500	45000	47500
% increase	--	5.8	11.76

Losses in motors

All motors have inherent losses. These are termed as fixed losses and variable losses. Fixed losses pertain to magnetic core losses (iron losses), friction and windage losses. Variable losses are due to current flowing through the windings (I^2R) and stray losses.

In standard motors such losses are stable initially but can deteriorate in use to factors mentioned above.

High efficiency motors

High efficiency in design is achieved by reduced losses

In high efficiency motors 2 to 5 percent additional efficiency is possible. For energy savings, it should be ensured that all replacements of existing motors are done with high efficiency motors. Additional cost for high efficiency motors is recovered in a short duration and thereafter savings of high efficiency and lower energy consumption continue.

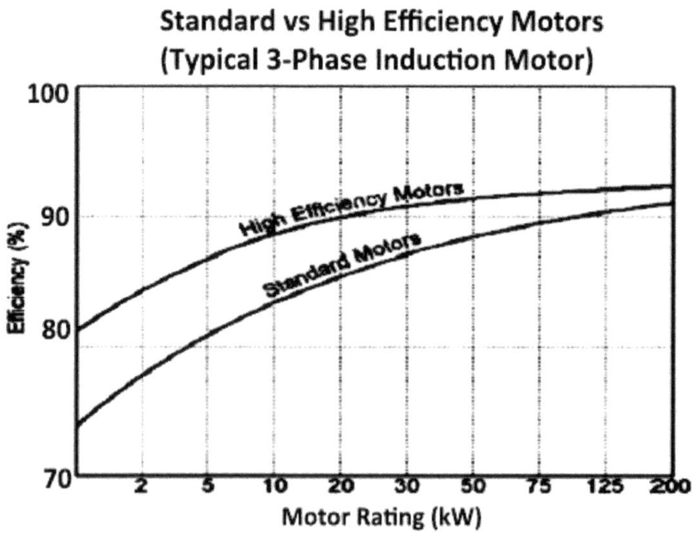

Courtesy BEE

Measurement of efficiency

For motors in operation, accurate measurement of efficiency is quite difficult and the plant management will have to resort to comparison or keep a check on deterioration by measuring current or energy consumption. Initially manufacture's test certificate is most relevant as that can show the no load current and KW etc. Subsequently, a plant which uses a large number of electric motors can have a motor audit conducted to know the accurate status of efficiency of motors.

Rewinding of motors

Every time a motor is rewound, there is some loss of efficiency. The loss level varies with quality of rewinding but it is not possible to restore the efficiency to its original level. Rewinding often involves burning of old winding, hammering to pull out bearings and other such actions which damages insulation and cause eccentricity in rotor and stator of the motor. Consequently, with repeated rewinding there is deterioration in efficiency and higher electricity consumption for the same output. Often the management only looks at capital expenditure in replacement ignoring the high consumption of energy. As a thumb rule a new replacement for rewound motor pays back in two years.

For a typical 10 HP motor, loss of efficiency can result in additional annual consumption of electricity which may be comparable to the cost of new motor. An example is given below.

Higher consumption can also be consequent to phase imbalance in motor windings. Incorrect voltage in operation can also cause higher energy consumption.

Rewound Motor

Performace of a rewound motor (On no load) Table Current (Average)

Parameter	Standard	Rewound
Voltage	415	415
Current (average	12.9	13.1
PF	0.2	0.31
kW	1.85	2.92
Excess loss	-	1.07 kW
Excess energy cost @ Rs. 3.5	-	Rs. 30, 000
Cost of new motor	Rs. 25,000	

Maintenance practice

Motors require very little maintenance but adhering to schedule provided by the supplier can ensure that good efficiency is continuously obtained in operation.

It is advisable to carry out on line monitoring in a plant if large number of motors are in use and effort is beyond periodic manual monitoring. On line monitoring can indicate when a motor is tending towards low efficiency and possible burnout.

Factors effecting energy efficiency of motors

Ideally, efficiency is the ratio of mechanical power delivered by the motor to the electrical power supplied. Factors effecting efficiency are as follows:

- Plants use mostly induction motors which have inherently low PF. For good efficiency PF is improved by providing PF capacitors. For large motors individually dedicated capacitors are used whereas for smaller motors upto 5 HP, these are grouped and capacitors for these motors are usually located at power centres or distribution boxes.
- Squirrel cage motors and large capacity motors are more energy efficient particularly those operating at higher speeds. This is because all techniques for energy efficiency are provided in such motors by the manufacturers.
- Motors operating at high temperature are less efficient. It is advisable to procure special motors for specific use in high temperature operation.
- Overloading or under loading of motors has adverse effect on their efficiency. Capacity of motor should be such as to provide for loading of 75 to 80% for required load. Often, consultants or designers have a tendency to play very safe and unnecessarily increase the motor capacity without sufficient reasons. Because of this reason, motor audits quite often indicate low level loading of around 30 to 40 %.
- Most of the motors in a plant are operated in part load. Initially, it should be ensured that motor HP or capacity is in conformance with its operational requirement. If for any reason it is known that part load operation for prolonged periods will be necessary then motors with good efficiency even at part load should be procured as part load performance of motors is usually less efficient unless the motors are so designed.
- Any deterioration in efficiency would be indicated if a record is maintained of no load power and current. Any increase in these parameters would be indicative of deterioration in its operation. Measurement of these parameters should be done accurately (preferably with the same/calibrated instrument) as change in value of current or power would be very small.
- Quality of input power to the motor is of much significance in motor efficiency as voltage fluctuation and frequency variation can reduce the running efficiency. Voltage is generally easier to improve and brought up to the required level. However, lower frequency cannot be controlled at the plant level if the supply is from external source like electricity board.
- Unbalanced voltage which implies that the voltage in three phases is not equal, adversely affects the efficiency. This unbalance in voltage is primarily due to different loading in three phases

- Good maintenance practices ensure that motors continue to operate at good efficiency. These relate to lubrication of bearings and their wear, alignment with load and check of electrical connections.

Motor standards

For the user industry, it is necessary to know the prevailing standards for electric motors. Globally, there is continuous effort to improve these standards and therefore the current standards should be kept track of. Indian standards have now been harmonized with international standards. Whenever new motors are ordered for replacement or new requirements are planned in projects, efforts should be made to procure motors of current standards so that energy efficiency is improved. It is possible to calculate the payback on additional cost incurred for a high efficiency motor through cost of reduced energy consumption. Motors are now available in following standards (BIS: IS 12615: 2011 and IEC 60034-1)

IE1 – Standard Efficiency: Comparable to **eff 2**

IE2 – High efficiency : Comparable to **eff1**

IE3 – Premium : Premium efficiency

IE4 – Super premium : Super premium efficiency

IE4 motors are not yet commonly used in India

4

Operational Efficiency of Pumps

All plants use pumps of different types – ranging from simple water pumps to more complex pumps for boilers, hydraulics etc.. However, not much attention is paid to the level of efficiency at which these are being operated.

Types of pumps

Pumps are designed to perform specific tasks and for pumping a specific fluid. A brief description of various types of pumps is given below to understand their best operation:

Reciprocating pumps: Also come under the category of positive displacement pumps. These are the oldest type of pumps and are still very common. Their main characteristics are low pumping volume and high head or pressure output.

Rotary pumps: These are most common form of pumps now and are characterized by high volume and low head or pressure. In this category there are various types of rotors such as centrifugal, vane type etc. Rotary pumps may also be classified as radial, axial etc.

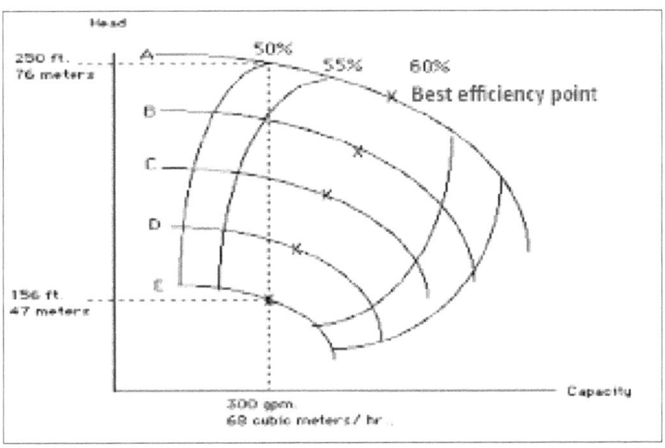

Courtesy BEE

Pumps can be operated singly or multi stage or in cascade for obtaining higher head or output pressure

Characteristic curve of pumps is given in the figure.

The above graph is drawn for a pump with capacity at x-axis and head at y-axis. Curves A B C D E represent various speeds at which the pump may be operated. Cross points x represent the best efficiency at the given speed. Most often, the pumps operate at a point other than at x which implies that the pumps operate at lower than the best efficiency possible. For a single stage pump, the efficiency is seldom above 60%. Thus it is necessary that by suitable changes to capacity and head, best possible efficiency is obtained.

In plants where there are many pumps in use and fluids are often pumped up or down, it is advisable to look at pumping of the fluid more holistically. Quite often the same fluid is pumped up to a high level and then this fluid is brought down to a lower level consuming energy both ways. A study to review the process technology for avoiding energy wastage is necessary. It is called ' pinch technique'.

Speed vs Power

% Speed	% Power
100	100
90	73
80	51
70	34
60	22
50	13

It can be seen from the above table that use of variable speed drives can be most effective for obtaining varying output from any pump. Any output control by throttling is inefficient and wastes energy. Similarly, for blowers use of dampers should be avoided.

Ways to get good efficiency

1. As stated above the pumps should be operated at as near the cross point as possible by making some changes in head, capacity(volume) and speed.
2. Throttling at output or flow control by use of valves should be avoided.
3. Where there is fluctuating output requirement, for better efficiency, variable speed drive should be used.
4. If the output is required to be lower for operational reasons one of the ways for achieving lower output and better efficiency is by reducing the impeller diameter.

5

Compressed Air System

Compressed air is most commonly used in all industries. It is also one of the most inefficient uses of energy in the industry.

Air is free but compressed air is not free

In the best of systems, energy efficiency of compressed air is in the range of 15 to 20 percent as substantial part of energy used for compression is wasted in heat generated in the process of compression. This heat is seldom usable and, on the other hand, compressed air may sometimes require to be cooled for its efficient use. The other problem with compressed air is leakages which are often allowed to linger and due to lack of awareness these are not immediately attended to. The table below indicates energy wasted in compressed air due to leakages.

Cost of air leakage

Orifice Size mm	KW wasted	*Energy Waste (Rs/Year)
0.8	0.2	8000
1.6	0.8	32000
3.1	3.0	120000
6.4	12.0	480000

** Based on tariff of Rs. 5 / kWh ; 8000 operating hours per annum; with pressure of 7.0 bar*

Source: BEE

Earlier, compressed air was commonly used in many production activities such as pneumatic tools, cooling and cleaning etc. However, these days use of compressed air is very restricted as industry has realized its low energy efficiency. It should also be understood that energy efficiency is best obtained when input air is at low temperature in a natural way which implies that air should not be cooled by expending energy as such. Air compressors should be located in cooler part of the plant and input air should also be taken from cooler location. Presence of moisture or humidity in input air is not desirable. In large compressors, energy efficiency can be improved through waste heat recovery.

When compressed air is distributed through a plant the network of distribution should be as short as feasible. For this purpose, compressors should be optimally placed and professional expertise sought to determine network size and compressor sizing.

Compressor efficiency

$$\text{Theoretical KW} = \left(NK / (K\text{-}1)\right)\left(Q \times P_s / 0.612\right)\left[\left(P_d / P_s\right)^{K\text{-}1/NK} - 1\right]$$

N	=	No. of stages
K	=	Ratio of specific heats (1.35 for air)
Ps	=	suction pressure in kg/cm2
Pd	=	Discharge pressure in kg/cm2
Q	=	Actual air flow (m3/min.)
Actual kW	=	$\sqrt{3}$ V I × PF as measured

Efficiency of an air compressor is usually expressed as power consumption in KW/ volume.

(Courtesy BEE)

For achieving optimum energy efficiency, a few practical steps listed below can be taken by plant personnel:

- Inlet air should be as cool as possible. This can be done by locating the air inlet of the compressor in cooler part of the plant.
- Air filters should be cleaned regularly
- Working pressure of compressed air should be set as low as possible considering its use and utilization. This is very important as there is a tendency in the industry to generate high pressure which may not be required at user end.
- At higher altitudes energy consumption for same outlet pressure is higher.
- In distribution of compressed air, pipe size should be optimum as narrow pipe can lead to higher energy consumption.
- To monitor the efficiency of compressor, observe change in duty cycle – load / unload – times. For the same air consumption at load end, if load duration increases, efficiency is declining.

- A situation where pressure of compressed air is throttled for lower pressure requirement, this results in avoidable wastage of energy. Use of variable speed drives should be considered.
- In the past, there was a tendency in some plants to use compressed air for cleaning purposes. Such misuse should be avoided.
- Use of pneumatic tools should be avoided unless necessitated by safety considerations. Similarly use of compressed air for mixing or stirring is not advisable.
- If requirement of pressure varies hugely then it may be better to use different compressors for different requirements in the plant.

Leakage test

It is advisable to conduct a plant wide leakage test periodically when the compressed air is not in use. For this test, compressor should be run to charge the system to upper pressure set. With no consumption in the machines, time taken for the pressure to reduce to lower limit for re-starting the compressor is indicative of leakage.

The system leakage is calculated as

System leakage (cmm) = Q ′ T / (T + t)

Q = Actual free air being supplied during trial, in cubic meters per minute

T = Time on load in minutes

t = Time unload in minutes

(Courtesy NPC)

Leak test: example

- Compressor capacity (CMM) = 35
- Cut in pressure kg/SQCMG = 6.8
- Cut out pressure kg/SQCMG = 7.5
- On load kW drawn = 188 kW
- Unload kW drawn = 54 kW
- Average 'On-load' time = 1.5 minutes
- Average 'Unload' time = 10.5 minutes

Comment on leakage quantity and avoidable loss of power due to air leakages.

a) $\text{Leakage quantity (CMM)} = \dfrac{(1.5)}{(1.5)+(10.5)} \times 35$

$= 4.375 \text{ CMM}$

b) Leakage per day = 6300 CM/day

c) Specific power for compressed air generation $= \dfrac{188 \text{ kWh}}{(35\times 60)\text{ CMH}}$

= 0.0895 kwh/m3

d) Power lost due to leakages/day = 563.85 kWh

Capacity Assessment in Shopfloor

- Isolate the compressor along with its individual receiver being taken for test from main compressed air system by tightly closing the isolation valve or blanking it, thus closing the receiver outlet.
- Open water drain valve and drain out water fully and empty the receiver and the pipe line. Make sure that water trap line is tightly closed once again to start the test.
- Start the compressor and activate the stop watch.
- Note the time taken to attain the normal operational pressure P2 (in the receiver) from initial pressure P1.
- Calculate the capacity as per the formulae given below :

Actual free air dischage $Q = \dfrac{P_2 - P_1}{P_0} \times \dfrac{V}{T} \text{NM}^3 / \text{Min}.$

Where

P2	=	Final pressure after filling (kg/cm^2 a)
P1	=	Initial pressure (kg/cm^2a) after bleeding
P0	=	Atmospheric Pressure (kg/cm^2a)
V	=	Storage volume in m^3 which includes receiver, after cooler, and delivery piping
T	=	Time take to build up pressure to P_2 in minutes

Example

- Piston displacement : 16.88 CMM
- Theoretical compressor capacity : 14.75 CMM @ 7 kg/SQCMG
- Compressor rated rpm 750 : Motor rated rpm : 1445
- Receiver Volume : 7.79 CM
- Additional hold up volume, i.e., pipe / water cooler, etc., is : 0.4974 CM
- Total volume : 8.322 CM
- Initial pressure P1 : 0.5 Kgf / SQCMG

- Final pressure P2 : 7.03 Kgf / SQCMG
- Atmospheric pressure P0 : 1.026 Kgf/cm2A
- Compressor output CMM : $\frac{(P_2 - P_1) \times \text{Total Volume}}{\text{Atm. Pressure} \times \text{Pumpup time}}$

$$\frac{(7.03 - 0.5) \times 8.322}{1.026 \times 4.021}$$

$$= 13.17 \text{ CMM}$$

Courtesy: NPC

If the readings of leakage test and capacity are regularly taken and recorded, these will provide invaluable information on condition of compressed air system.

Application of types of compressors

Reciprocating Compressors: These are most widely used as it is possible to obtain high pressure output in single stage. Suitable for high pressure at moderate or low output.

Rotary compressors: These provide a continuous output and can be operated at high speed. Suitable for high output at moderate pressure. Rotary compressors can be of different types such as centrifugal or screw type.

Centrifugal compressors are suitable for requirements of very high capacities of compressed air. These compressors are oil free and thus suitable for such applications.

Dust free air intake

By air filtration at the input the efficiency of compressor and equipment being supplied with compressed air increases. Air filters should kept clean by undertaking periodic cleaning.

Dry air

Dry air at inlet is good for energy efficiency. Air intake should accordingly be taken from dry area.

For some applications dry air free of moisture is required. This is done at some cost to energy efficiency by means of refrigeration or absorption method

Pressure setting

Proper setting of output air pressure is very significant for energy efficiency of the air compressor. It is observed that output pressure is set much higher than working pressure required at the load end.

To regulate pressure during heavy or light loading air receiver is used. Receiver is provided with cut off and cutting in arrangement. This setting translates into load and unload cycle of the air compressor and hence the efficiency. If the setting at the unload cycle is unnecessarily high then load cycle of the compressor will commence earlier than needed resulting in higher energy consumption. If the delivery pressure is lowered by 1 bar, energy consumption may reduce by 6 to 10 %.

Use of ring main for compressed air is desirable.

In conclusion, it can be stated that use of compressed air should be minimized as its production is very inefficient.

6

Boilers and Steam Distribution

Boilers are very widely used in industry and they offer maximum scope for energy efficiency and savings in fuel. However, these are seldom given the attention for either energy efficiency or fuel savings. Perhaps because boilers are only routinely maintained by staff at lower level who do not get involved in energy conservation practices in the plant.

Boilers are unique in this respect that there is specific law and regulation namely Indian Boiler Regulation which applies to all boilers having a capacity of 22.75 litres and above. This also applies to pipes carrying steam at pressure 3.5 Kg/sq cm or exceeding internal diameter exceeding 254 mm.

In the case of boilers, efficiency deteriorates with time due to poor combustion, loss in insulation, scaling and other maintenance problems which are not immediately discernible. The minimum effort required with boilers is to measure their efficiency periodically. If the reduced efficiency is observed, it can enable measures to locate the deficiency and correct it.

Measurement of boiler efficiency

There are mainly two ways of measuring the efficiency of boilers: 1. Direct and 2. Indirect

For direct efficiency

$$\text{Boiler efficiency } (\eta): = \frac{Q \times (H - h)}{(q \times GCV)}$$

Where

Q = Quantity of steam generated per hour (kg/hr)

H = Enthalpy of saturated steam (kcal/kg)

h = Enthalpy of feed water (kcal/kg)

q = Quantity of fuel used per hour (kg/hr)

GCV = Gross calorific value of the fuel (kcal/kg)

Direct efficiency has advantages such as quick evaluation, easy computation and fewer instruments for monitoring. Disadvantages are mainly because of difficulty in measuring quantity of steam generated, lack of analysis and discrimination in losses.

For indirect efficiency measurement losses in a boiler are measured by flue gas analyser. As given in the diagrams below, losses measured by a flue gas analyser are deducted from the input fuel to arrive at the output for efficiency calculation.

Specific consumption

12.5 to 13 kg of steam /kg of HSD/LDO

- 10.8 to11.3 kg of steam/ Lit of HSD/LDO

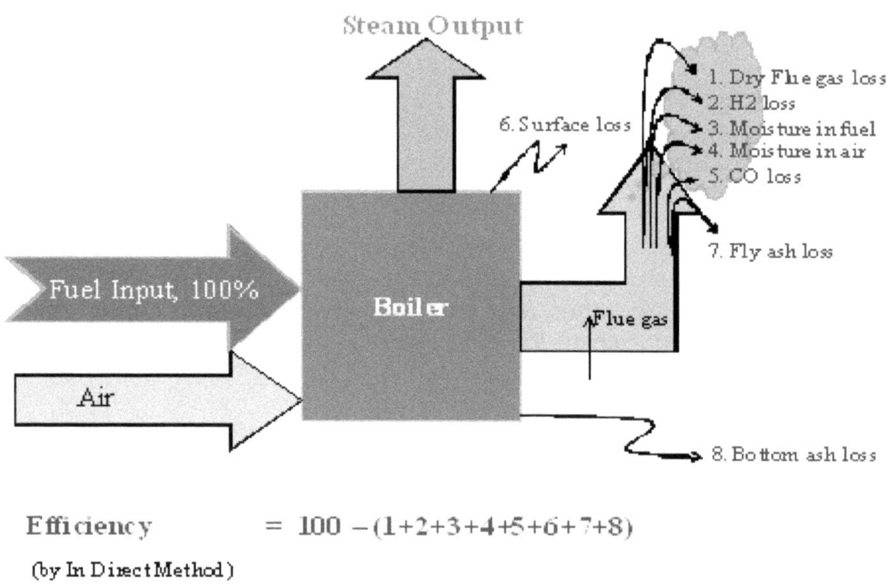

Courtesy BEE

Input/Output Parameter		Kcal / Kg of fuel	%
Heat Input in fuel	=		100
Various Heat losses in boiler			
1. Dry flue gas loss	=		

Input/Output Parameter		Kcal / Kg of fuel	%
2. Loss due to hydrogen in fuel			
3. Loss due to moisture in fuel	=		
4. Loss due to moisture in air	=		
5. Partial combustion of C to CO	=		
6. Surface heat losses	=		
7. Loss due to Unburnt in fly ash	=		
8. Loss due to Unburnt in bottom ash	=		
Total Losses	=		
Boiler efficiency = 100 – (1+2+3+4+5+6+7+8)			

As shown above losses are due to dry flue gas, moisture in fuel and air, combustion of hydrogen, heat radiated from boiler and due to unburnt fuel

By indirect method, efficiency of Boiler (h) = 100 – (1+2+3+4+5+6+7+8)

Both methods offer some challenges: For direct measurement, quantity of steam is required which is difficult to measure. However, fairly accurate steam flow meters are now available to provide this parameter. In the case of indirect method, depending upon flue gas analyser and radiation/convection losses (estimated as 1 to 2 %), which are difficult to assess, accuracy of efficiency figure may suffer while calculating the losses.

Maximising Energy Efficiency

1. Stack temperature: It should be as low as possible. At the same time if it is too low, exhaust may condense on walls of stack and it has a tendency to settle towards ground after leaving the stack. This is not acceptable so the stack temperature should be kept slightly above 200°C. Higher stack temperature indicates scope for heat recovery.
2. Use of Economiser: Most of the current boilers use flue gases from the boiler for pre heating of feed water to the boiler. This is done by using economizer. If the boiler is not fitted with economizer it should be installed as early as feasible.
3. Pre-heating of air: Air supplied for combustion in boiler could be pre heated using flue gases.
4. Combustion: For obtaining maximum energy efficiency it is important that fuel is fully burnt. Incomplete combustion can be due to less air or irregular supply of fuel. In case of coal, size of lumps can cause improper combustion though the usual method for large boilers is to pulverise coal. On the other hand fine particles may cause obstruction of air.

5. Excess air: For complete combustion, air supplied to a boiler is higher than chemical requirement by approximately 10% depending on fuel. If the excess air is more than optimum level, then it can result in loss of efficiency.

6. Loss of heat due to radiation and convection: Heat loss takes place through boiler shell and may be of the order of 1 to 2%. Loss can be minimized by having good insulation.

7. Blowdown: Steam blowdown to reduce air from ingress can be achieved by ensuring that blowdown takes place regularly. If the interval is too short or too long then EE can get affected. Automatic blowdown system is preferable,

8. Scaling and soot on tubes: Deposits of scaling and soot need to be removed periodically as these inhibit transfer of heat. High exit temperature of gas indicates poor heat transfer. There is an estimated loss of 1% efficiency when the stack temperature rises every 22°C.

9. Boiler steam pressure: This is a common occurrence in the industry that steam pressure generated is much higher than necessary for working pressure required in distribution. Generally it is kept at the highest pressure required for any process or for various machines installed. By generating steam at lower pressure, better efficiency can be obtained.

10. Using VSDs: Use of VSDs, for various auxiliaries like pumps instead of throttling of the boiler can save energy which is basically supplied from the boiler.

11. Operation of Boiler at maximum efficiency: Maximum efficiency is achieved at approximately two third loading of boiler capacity.

Best practices for steam distribution

1. Steam mains should be run with a falling slope (125 mm for every 30 meters) in the direction of steam flow.

2. Drain points should be provided every 30 to 45 meters. Drain pipes are also necessary at low points in steam pipes, reducing valves and stop valves.

3. Piping should have steam traps at suitable points which must be kept functional.

4. Branch lines from the mains should be connected at the top as otherwise these can become drains for the main line.

Steam Distribution – Steps for achieving good energy efficiency

1. Leakages: One of the basic and obvious factors is avoidance of steam leakage. It is important as there is loss of efficiency due to steam leakage. It is estimated that a 3mm hole in a pipeline with steam pressure 7Kg/sqcm^2,

would be wasting 33 KL of fuel oil annually. It is not uncommon in plants having steam distribution that steam leakages are visible externally.

It is advisable to replace rarely opened flanges in a pipe by welded joint.

2. Dry Steam: At the delivery point for a process or a machine, steam should be dry (not wet) as otherwise corrosion can occur. Wet steam reduces total heat in the steam. Dry saturated steam is better than superheated steam for industrial heating.
3. Pressure: Steam pressure should be set as required in the process. For process equipment, dry saturated steam is necessary and it should be utilized at lowest possible pressure. It is often seen that steam is produced at high pressure and then used at lower pressure by using reducer.
4. Insulation: Insulation of steam pipes should be checked periodically and replaced when necessary. This will include the process or machinery where the steam is being supplied.
5. Steam traps: These are necessary in any steam distribution system but after installation these are often neglected. Steam traps are devices which discharge condensate and air from the line or equipment, but retain steam. Depending upon the type of steam trap used its effective functioning should be monitored regularly.
6. Blowdown: Periodically the boilers are blown down to control the level of dissolved solids. Blowdown helps to protect the surface of heat exchangers among others.
7. Proper Air Venting: The presence of air inside the equipment will reduce the partial pressure of steam in the steam air mixture. Adequate air venting should be made at appropriate positions in the pipelines to purge out air from the equipment as quickly as possible to make the heat transfer more efficient.
8. Condensate Recovery: Steam condenses after discharging its latent heat but still retains a sizable portion of heat. It is advisable to recover the condensate and use it for boiler feeding and reduce the fuel requirement.

Thus, it can be seen that steps outline above may be somewhat cumbersome but enhancing efficiency of boiler results in considerable saving in expenditure of steam generation.

7

Efficiency of Fans and Blowers

All plants have fans and blowers to meet the requirements of air for process, cooling etc. The efficiency of fans and blowers is very important as these are running continuously. Classification of fans, blowers and compressors depends upon the compression ratio or rise in pressure from inlet to outlet.

Comparison of fan efficiency: Selection of fan would depend on volume requirement, rate of flow, pressure and physical fitment. The efficiency of various type of fans varies from 50% to 80%.Blowers on the other hand can provide higher pressure. Other factors which effect efficiency are as follows:

System characteristics: Resistance to flow of air is the main criterion as system resistance varies squarely with the volume of air flow. If the duct is narrow or has too many bends the efficiency would be lower due to high resistance.

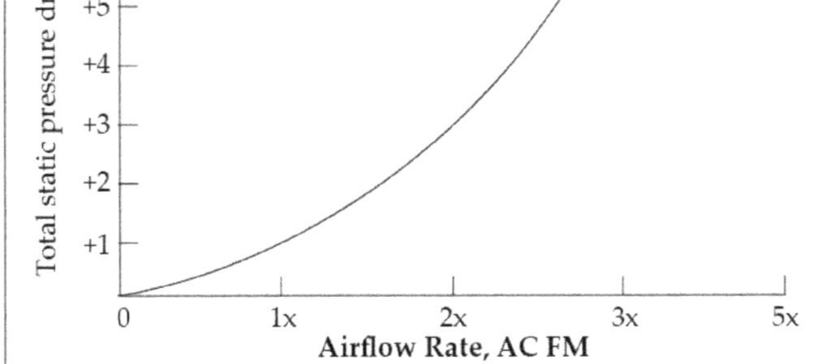

Courtesy BEE

Characteristics of fan: Performance of a fan depends on volume, system pressure, fan speed

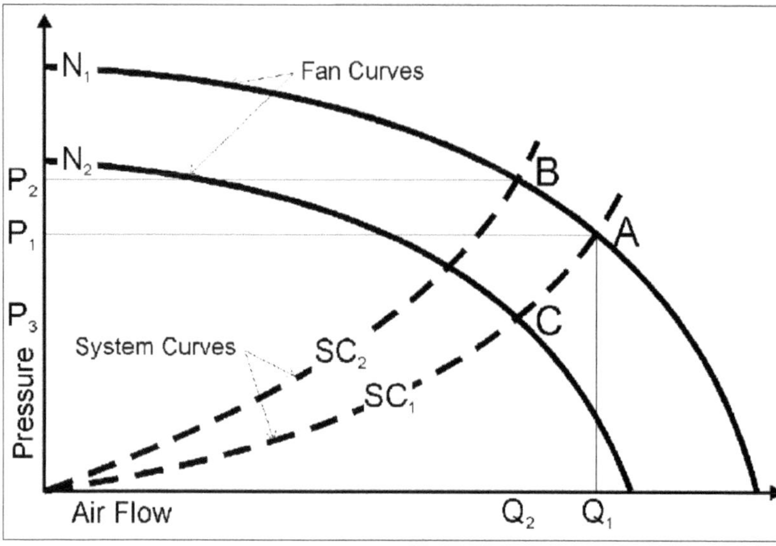

Courtesy BEE/NPC

Some factors governing Fans

For best efficiency in fans, it is essential to know the correlation of speed, power and pressure. Change in speed results in change of pressure rise and power consumption.

Varying the speed (RPM) by 10% would change the air delivery also by 10%. Thus,

$$Q_1/Q_2 = N_1/N_2$$

Varying the speed by 10% would change the pressure by square of the speed viz 19%

$$P_1/P_2 = \{ N_1/N_2 \}^2$$

Varying the speed by 10% would change the power consumption by cube of the speed viz 27%

$$KW_1/KW_2 = \{ N_1/N_2 \}^3$$

Ways to achieve Energy Efficiency

1. Use of high efficiency impeller and high efficiency fan motor/assembly
2. Replace output throttling or discharge damper and instead reduce impeller diameter.
3. Use of VSD for speed control.
4. Minimise demand on the fan by proper control of excess air, air leaks etc.

8

Transformers and Their Efficiency

Transformers are very efficient devices and keep operating for over two decades without breakdown. An ideal transformer is that which has no losses. In practical terms, transformer losses are always present but comparatively very small losses exist but these remain present throughout their long running lives. Therefore, it is very important to understand transformer losses and carry out proper selection at the initial stage itself.

Losses in transformers

There are basically two types of losses in transformers:

a) Iron losses or No load losses: These losses are caused in the core and are constant through the life of the transformer.

b) Copper losses or load losses: These are caused by current flowing through transformer winding and vary with loading of transformers.

Total losses P = P (No load) + (% load) ^{2} X P(Load)

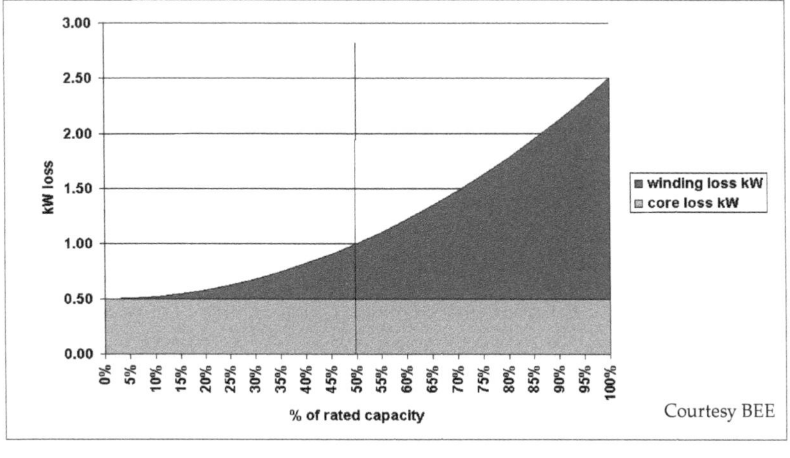

Courtesy BEE

Industrial Transformers

While there are many types of transformers, those in use in industry are mainly for converting electricity board supply at high voltage to voltage required for distribution in the plant. Such transformers are similar to distribution transformers used by electricity boards.

BIS Standards relevant are:

IS 1180 (Part 1) 1989: Outdoor type 3 phase upto 100 KVA 11KV Non sealed
IS 1180 (Part 2) 1989: Outdoor type 3 phase upto 100 KVA 11KV Sealed type
Revision IS 1180 Outdoor type insulated liquid immersed upto 2500 KVA 33 KV

In BIS, standard ratings, insulation levels, No load voltage ratios, Maximum total losses, limits of temperature rise among others are specified.

Fault detection

Large transformers are supported with means to detect and analyse gases from the transformer. These gases can be evaluated for fault detection:

- If there is arcing, hydrogen and acetylene in substantial quantities may be detected.
- Methane indicates local heating and/or hot spots
- Overheating of insulation can result in carbon monoxide and dioxide.

Condition monitoring

As already mentioned the transformers are very rugged and require little maintenance. However, condition monitoring which includes check of insulation resistance of windings, check on winding and oil temperature , is necessary.

Condition monitoring helps in:

- Availability of relevant data
- Detection of gases and faults indicated
- Timely attention to correctives and changes necessitated in scheduled maintenance.
- Avoidance of major failures
- Draws attention to safety aspects

Causes of failure

Transformers are quite robust and do not often fail if normal maintenance is carried out. However, under mentioned causes may result in failure:

- Contamination of insulation and ageing of insulation in winding
- Short circuit causing winding deformation
- Core insulation failure
- Overvoltage and overheating

9

Refrigeration and Air Conditioning Efficiency

Many process industries use refrigeration in their production. Also many production areas are air conditioned. Since energy in large amounts is used in refrigeration and air conditioning, both offer immense scope for energy saving.

Refrigeration is involved where the material or process requires cooling and for best efficiency such material should be at optimum temperature prior to commencement of cooling. In case of air conditioning, there are basically three sources of heating in the production area. These are inward flow of heat from outdoors at higher temperature, internal heat due to equipment or process and persons, and direct heat from sun.

For best efficiency in air conditioning, all three sources should be controlled by elimination, insulation and reducing ingress

Cooling effect

As refrigeration and air conditioning are based on the same technique, cooling effect is quantified as tons of refrigeration which also defines the capacity of refrigeration or cooling system. In refrigeration, one ton is equal to 3024 kcal/hour of heat rejected. A typical air conditioning system is shown below:

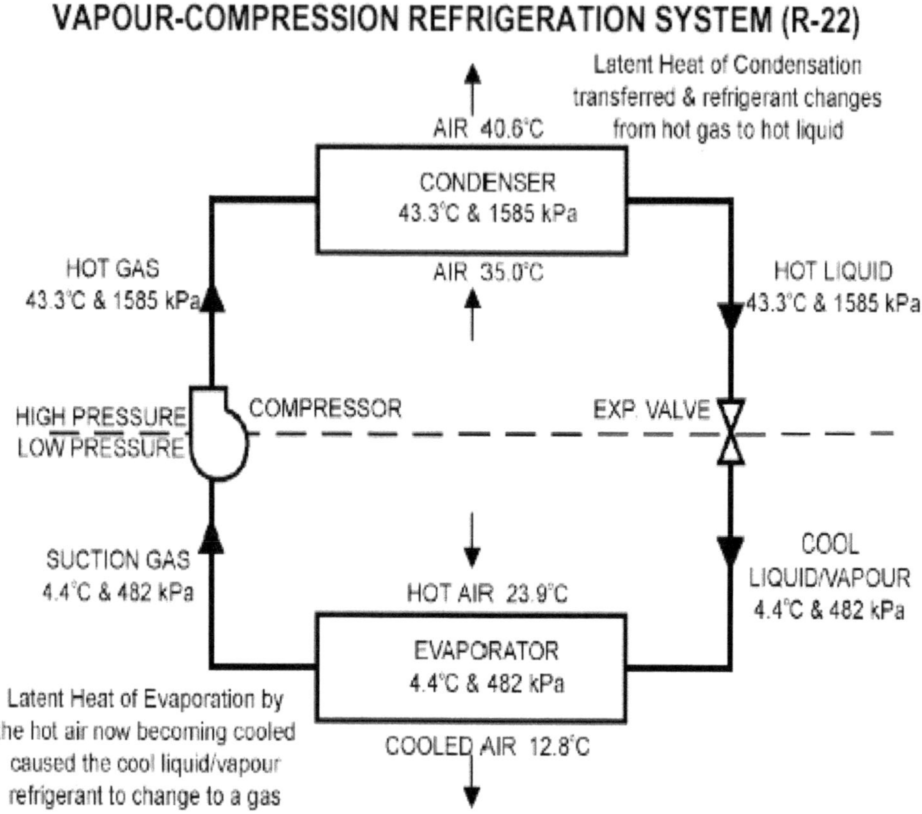

Courtesy: BEE

The refrigerant is compressed in the compressor reaching a high temperature and sent to condenser. In the condenser, the refrigerant is cooled and sent to expansion valve where the high pressure and low volume at the inlet is reduced to low pressure and high volume bringing down the temperature. Low temperature refrigerant is sent to evaporator which is located where the required space is to be cooled. After cooling the load and absorbing the heat from space, the refrigerant at higher temperature is sent to compressor completing the cycle of conventional refrigeration aka vapour compression system. This system is most commonly used in small window type to very large refrigeration or air conditioning systems.The compressor in the system may be reciprocating type, centrifugal or screw type in the same way as other compressors.

The other system for refrigeration or air conditioning is **Vapour Absorption system.** The absorption refrigeration system differs fundamentally from vapour compression system only in the method of compressing the refrigerant. An absorber, generator and pump in the absorption refrigerating system replace the compressor of a vapour compression system.

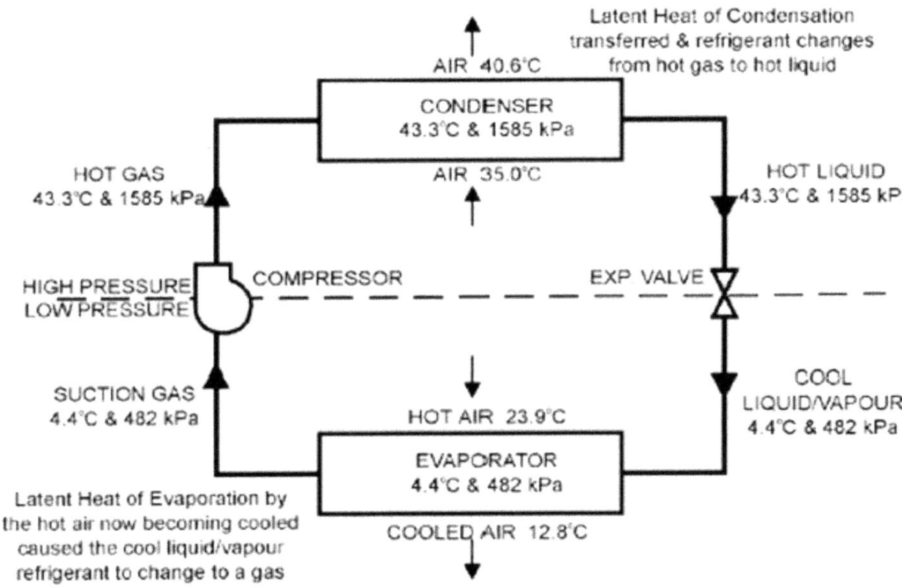

Courtesy BEE

As shown above, Ammonia vapor is produced in the generator at high pressure from the strong solution of NH3 by an external heating source. The water vapor carried with ammonia is removed in the rectifier and only the dehydrated ammonia gas enters into the condenser. High pressure NH3 vapor is condensed in the condenser. The cooled NH3 solution is passed through a throttle valve and the pressure and temperature of the refrigerant are reduced below the temperature to be maintained in the evaporator. The low temperature refrigerant enters the evaporator and absorbs the required heat from the evaporator and leaves the evaporator as saturated vapor. Slightly superheated, low pressure NH3 vapor is absorbed by the weak solution of NH3 which is sprayed in the absorber Weak NH3 solution (aqua–ammonia) entering the absorber becomes strong solution after absorbing NH3 vapor and then it is pumped to the generator through the heat exchanger. The pump increases the pressure of the strong solution to generator pressure. The strong NH3 solution coming from the absorber absorbs heat from high temperature weak NH3 solution in the heat exchanger. The solution in the generator becomes weak as NH3 vapor comes out of it. The weak high temperature ammonia solution from the generator is passed to the heat exchanger through the throttle valve. The pressure of the liquid is reduced to the absorber pressure by the throttle valve.

Refrigeration Performance

The specific power consumption kW/TR is a useful indicator of the performance of refrigeration system. By measuring refrigeration duty performed in TR and the

Kilo Watt inputs in electrical power, the ratio kW/TR is used as a reference energy performance indicator.

The refrigeration TR is assessed as TR = Q $\times C_p \times (T_i - T_o)$ / 3024
Where TR is cooling TR duty
Q is mass flow rate of coolant in kg/hr
C_p is coolant specific heat in kCal /kg / ^{0}C
T_i is inlet. Temperature of coolant to evaporator (chiller) in °C.
T_o is outlet temperature of coolant from evaporator (chiller) in °C.

The theoretical Coefficient of Performance (Carnot), $COP_{Carnont}$ - a standard measure of refrigeration efficiency of an ideal refrigeration system- depends on two key system temperatures, namely, evaporator temperature T_e and condenser temperature T_c with COP being given as:

$$COP_{Carnot} = T_e / T_c - T_e$$

This expression also indicates that higher COP_{Carnot} is achieved with higher evaporator temperature and lower condenser temperature.

COP_{Carnot} is only a ratio of temperatures, and does not take into account the type of compressor. Hence the COP normally used in the industry is given by

$$COP = \text{Cooling Effect (kW)}$$

Power input to compressor (kW)

where the cooling effect is the difference in enthalpy across the evaporator and expressed as kW.

Overall energy consumption is total of
Compressor kW
Chilled water pump kW
Condenser water pump kW
Cooling tower fan kW
Overall kW/TR = sum of all above kW/ TR

Effect of Poor Maintenance on Compressor Power Consumption

Condition	Evap, Temp (°Temp)	Cond Temp (°C)	Refrigeration Capacity	Specific Power Consumption (kW/ton)	Increase in kW/ Ton (%)
Normal	7.2	40.5	17.0	0.69	-
Dirty condenser	7.2	46.1	15.6	0.54	20.4
Dirty evaporator	1.7	40.5	13.8	0.82	18.3
Dirty condenser and evaporator	1.7	46.1	12.7	0.96	18.3

Comparative performance of different types of compressors in air conditioning

Centrifugal compressors: Best efficiency is obtained when centrifugal compressors are operating near full load. Their main advantage is efficiency in large sizes and therefore their application is in large chillers. At reduced loads, efficiency is lower as centrifugal compressors output is throttled by inlet vanes at inlet to impeller. Use of VSD is a better alternative.

Reciprocating compressors: At maximum, the efficiency of reciprocating compressors is below that of centrifugal or screw compressors. However, at partial loads there is minimum loss of efficiency. Smaller reciprocating compressors can be turned on/off. VSDs can be used if necessary.

Screw compressors: Generally preferred for higher condensing pressures and are quite compact. These are amenable to variable speed drives.

Air handling units(AHUs) and Fan Coil units (FCUs)

AHUs consist of cooling coils and fans to supply cooled air to an area. AHU may have either fresh air or mix of fresh and return air. AHU is supplied with chilled water from central chiller with its temperature control for valves supplying air to cooled space.

FCU is a small terminal device (compared to a more elaborate AHU)which has a cooling coil and fan supplied with chilled water. It has a control valve depending on room temperature.

Energy saving measures in refrigeration and air conditioning

Low temperature requires more energy. Therefore, it is advisable to maintain comfort conditions: 25°C, 55 % RH

Minimize ingress of heat load through glass windows; provide sun control film, use double glass

Insulate roof top in A/C Building, use white paint and /or plants

Insulation of cold vessels and pipelines is necessary to obtain good refrigeration efficiency

Provide under deck insulation of 50 mm,

Optimize fresh air supply into a/c room

Conduct CO_2 study to optimize fresh air quantity(10-15 cfm/person or 0.25 cfm/sq..ft as per ASHRAE)

Minimise artificial lighting; use natural lighting , 3.5 kw lighting consumes 1.0 TR load

Provide controls; install thermostat to control peak and base load

Provide VSD for AHU with return air temp. sensor set at 25°C

Air tight the building envelop; prevent cold air leakage, Provide door closures

Avoid heat producing equipments inside the room; keep away UPS Battery , ovens, other loads

10

Cooling Towers

Cooling towers are an essential feature of most of the industries. However, cooling towers do not get the required attention as most of these are located away from the main plant machinery.

Factors for efficient operation of cooling towers

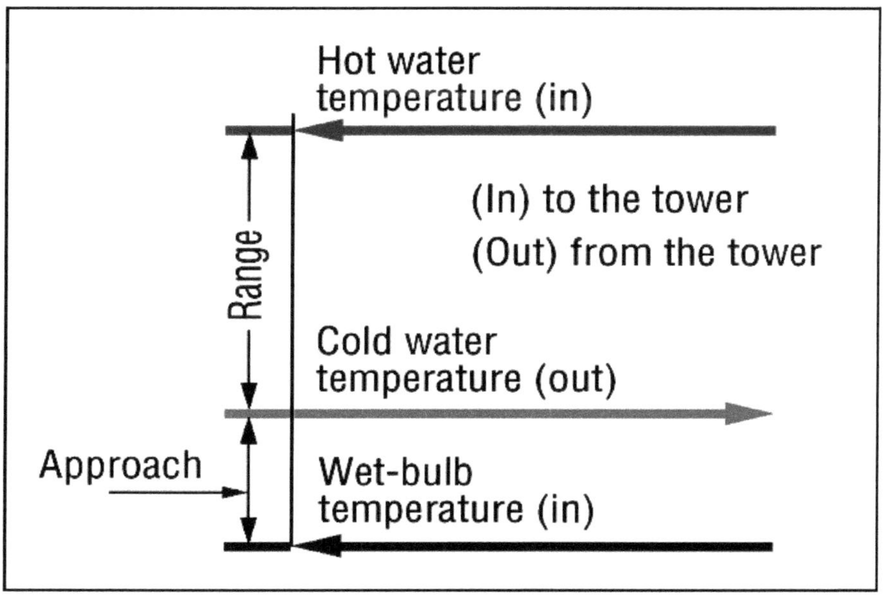

Courtesy BEE

The factors are dependent on temperatures of inward hot water, outgoing cooled water and wet bulb temperature which is the minimum possible temperature at which cooling tower can cool. Thus, these factors are easily determined with a thermometer.

Range: It is the difference between input hot water and output cooled water. Design, size and structure of cooling tower depend on the range desired.

Approach: It is the difference between temperature of wet bulb thermometer and the output cooled water. It is desirable that this difference is minimum and approximately 3 to 5 degrees Celsius. Cooling towers are usually designed for this. If this factor is higher it would indicate that cooling tower is not operating efficiently.

Ways to achieve efficiency in cooling towers

1. There should be free flow of air around cooling tower as any obstruction in the form of trees; buildings etc can reduce its efficiency.
2. Nozzles in the cooling tower should be clean or non clogging ones should be used.
3. As far as possible, incoming hot water should not be at very different temperatures and the source should be similar or homogenous. Segregation of high heat loads like furnaces, air compressors, DG sets should be separated from air conditioner plant or condenser
4. If the blade angle of fan is adjustable it should be according to season or load basis. FRP blades are more efficienct.

11

Fuels and Combustion Efficiency

Almost all industries use some fuels for production activities. The fuels are high value items and have a key role in economy of manufacturing. Any saving in fuel burning contributes directly to economic production as well as energy efficiency.

Characteristics

There are many characteristics of fuels which are important and have a bearing on their usage. These are specific gravity, viscosity, specific heat, pour point, flash point and calorific value. For industrial use, calorific value is most significant. Other factors are sulphur content, ash content or water vapour which tend to diminish the efficacy of fuel. For fuels, other considerations are availability, storage, handling, pollution and cost. All these need to be considered for selecting the appropriate fuel for a given industry and its location.

Calorific value is the measure of heat or energy produced and specified as Gross Calorific Value (GCV) or Net Calorific Value (NCV). The difference is the latent heat of water vapour produced in combustion. Most of the liquid fuels in common use such as Kerosene, furnace oil, diesel, LDO, LSHS etc have a calorific value between 10,000 to 11,000 Kilo calories per kilogram whereas Indian coal has calorific value near about 4000 kilo calories.

Certain liquid fuels such as furnace oil have a low pour point as they are in semi solid state at room temperature. Due to this reason they require streaming or heating of pipes to facilitate the flow. Furnace oil is still preferred for economic reasons.

Specific heat of fuel helps to quantify steam or electrical energy required for preheating of the fuel.

Proximate analysis in coal indicates the percentage by weight of carbon, volatiles, ash and moisture content. While carbon content and volatiles directly contribute to heating value, ash content and moisture are inhibiting factors. However, these have to be taken into consideration when designing fuel delivery, handling and disposal system.

Ultimate analysis indicates the various chemical constituents in fuel for determining the air required for complete combustion. (Note: For complete combustion certain amount of excess air is required over theoretical which is termed as **Stoichiometric** air)

Gaseous Fuels

In common use the gaseous fuels are LPG(Liquefied Petroleum Gas), CNG (Compressed Natural Gas), LNG (Liquefied Natural Gas) among others.

LPG is mixture of propane and butane with some other unsaturated fractions. It is denser than air and tends to settle down to the lowest level. Small quantities can give large volume of vapour with air and can be hazardous. LPG is deliberately odorized to help in leak detection.

Natural Gas is mainly methane and has high calorific value. It is lighter than air and if leaked it disperses easily into air.

Combustion

Combustion is rapid oxidation of fuel which is accompanied by production of heat and/or light. For complete combustion adequate amount of oxygen is necessary. In atmosphere nitrogen is the dilutant and is always present with oxygen. On combustion, nitrogen absorbs heat.

There are three **T**s for control of combustion – temperature, turbulence and time. Temperature should be high enough to cause combustion, turbulence refers to mixing of fuel and oxygen and time should be sufficient for complete combustion.

For optimum air supply the draft system would match with air requirement. For smaller or simpler requirement natural draft would be adequate whereas with increasing size forced or induced draft would be necessary.

For coal, the air requirement would be 7 to 8 Kgs for every Kg of coal. This is theoretical or Stoichiometric air. Depending on the system, approximately 10 % excess air would be required for complete combustion.

For oil, efficiency would be possible by atomizing the fuel so that complete combustion takes place.

For gases, the complete combustion may be achieved with a ratio of 9.5:1 or 10:1

Other considerations for efficiency

Pumping: Fuels should be pumped with positive displacement pumps as these can move the fuel even when it is cold. Centrifugal pumps are not suitable except for light fuels. At lower temperatures viscosity of heavier fuels increases making the pumping inefficient unless pre-heating is resorted to.

Sizing of coal: Proper size of coal ensures efficient combustion. Proper sizing relates to type of firing system so that there is even burning, reduced ash losses and better combustion. Coal is reduced in size by crushing and pulverizing. Screens are used to separate fines and small particles avoiding extra fineness. Magnetic separators are used for keeping out iron pieces.

Combustion controls: For best efficiency control of fuel supply, air supply and removal of exhaust gases is necessary. Where steam consumption varies these controls can help in ensuring optimum fuel efficiency. The controls would be more elaborate with size so that there is continuous optimization of fuel and air.

12

Harmonics and Power Quality

In the industry, not much importance is attached to harmonics and power quality in spite of these having huge adverse effects in certain type of manufacturing practices. Electricity boards also have no regulations for harmonics but are only concerned with power factor which is a part of power quality. It is ironic that certain energy efficiency measures have adverse effects on harmonics and power quality. Consequently, this aspect of energy efficiency is neglected most often and even the industry where harmful effects take place there is lack of awareness of 'side effects' of energy efficiency devices. In the industry, mitigation of ill effects of harmonics and power quality is usually not carried out.

What are harmonics?

Ideally, alternating current consists of true sinusoidal wave at fundamental frequency which should flow through electricity distribution in industry and its equipment. However, many equipments including motors and their adjuncts like variable speed devices, electronics including rectifiers, inverters etc, capacitors and inductors cause distortion of true sinusoid in current. As explained in Fourier transforms, combination of various sinusoid frequencies (integer multiples of fundamental frequency) result in distorted waveform. Conversely, a distorted waveform comprises a number of sinusoids as integer multiples of fundamental, of various frequencies, amplitudes and phase angles.

Distortion occurs in voltage as well as current.

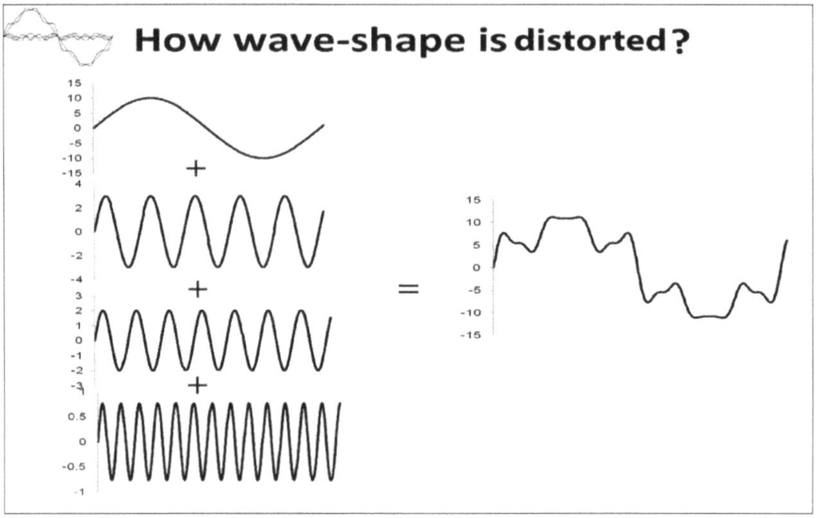

Courtesy Schneider Electric

Harmful Effects of Harmonics

Some of the harmful effects of harmonics are given below. It should be understood that in any location all the effects will not occur.

- Poor power factor which is not responsive to correction
- Unexplainable overheating of motors and transformers
- Failure of capacitors and nuisance tripping of contactors
- Increase in KVA demand

Harmonics Distortion Standards

There is no BIS or Indian standard for Harmonics distortion. IEEE 519 lays down current and voltage distortion standards

Current Distortion Limits

I^{sc}/I^{1}	<11	11<h<17	17<h<23	23<h<35	35<h	THD
<20	4.0	2.0	1.5	0.6	0.3	5.0
20<50	7.0	3.5	2.5	1.0	0.5	8.0
50<100	10.00	4.5	4.0	1.5	0.7	12.0
100<1000	12.0	5.5	5.0	2.0	1.0	15.0
>1000	15.0	7.	6.0	2.5	1.4	20.0

Courtesy Schneider Electric

Voltage Distortion Limits

Bus Voltage at PCC	Individual Voltage Distortion	THD
<69 kv	3.0	5
69>161 kv	1.5	2.5
>161 kv	1.0	1.5

Abbreviations:

Isc – Short circuit current for the facility

I^1 - Fundamental current

PCC – Point of common coupling

THD –Total Harmonic Distortion

There is no correlation between distortion amplitude and harmful effect. Similarly, it is not possible to predict the effect of voltage distortion on generation of current distortion.

Commonly noticed effects of Harmonics

- Blinking of Incandescent Lights – Transformer Saturation
- Capacitor Failure – Harmonic Resonance
- Circuit Breakers Tripping – Inductive Heating and Overload
- Computer Malfunction – Voltage Distortion
- Conductor Failure – Inductive Heating
- Electronic Equipment Shutting down – Voltage Distortion
- Flickering of Fluorescent Lights – Transformer Saturation
- Fuses Blowing for No Apparent Reason – Inductive Heating and Overload
- Motor Failures (overheating) – Voltage Drop
- Neutral Conductor and Terminal Failures – Additive Triplen Currents
- Electromagnetic Load Failures – Inductive Heating

Harmonic Analysis

If any of the symptoms are observed in a facility, the best way to address this problem is to engage a professional agency to conduct Harmonic Analysis. With that, it will be possible to get information on extent and magnitude of harmonics existing. Subsequently, appropriate solution for mitigation of harmonics can be determined.

Methods of improving Power factor

Fixed compensation is carried out for steady loads, no load compensation of motors, no load compensation for transformers and similar requirements. For varying load, varying compensation is the answer.

Type of compensation

Individual compensation is directly provided at loads such as motors. Group compensation is more economical when, for example, a number of motors are being fed from one supply point- distribution box, breaker etc.

Central compensation is carried out at the incomer. This enables avoidance of any penalty by the electricity board but it does not stop internal losses in distribution in the facility. Use of automatic power factor controller is quite common.

13

Efficiency of Secondary Generation

With the state of power generation and distribution, all industries are compelled to install standby generation capacity to operate all or essential equipment during outages. This also becomes necessary due to non availability of stable voltage in power supply. When there is shortage of power supply with the utilities, industries get low priority as compared to domestic use in spite of higher tariff paid by industry. As outages can last for several hours and even days, it is necessary to ensure that standby generation is efficient.

The plants which have steam in use can install co-generation based on low pressure steam available after usage in the process or from last stage of steam turbine.

Efficiency of conventional diesel generators

Diesel generator is the most common in standby mode in the industries. A typical diesel generator used in industry gives an efficiency of 35% or less due to which power from such generators costs very high when compared with grid supply. One litre of diesel generates 3.5 to 4.5 units of electricity depending upon the efficiency of generator. The price diesel available to the plant can be used to calculate the cost of power generated. As compared to grid power or supply from the electricity boards, the cost of power generated from a diesel generator is very high – 2to 3 times. Consequently, efficiency of generating this power is very important.

Factors applicable for good efficiency in DG sets

1. Cooling: DG sets may be air cooled in lower capacities and are usually water cooled for higher ratings. For air cooled sets, cross ventilation should be available for best results.
2. Power factor: For load on DG sets also good PF should be maintained for efficiency so that KVA load is not high when PF is low. Sometimes, manufacturers advise against using PF capacitors because their DGs are ingeniously rated – for example a DG may be rated as 1000KVA DG at 0.8

PF which implies that it is good only for 800 KW. Even after considering their rating, as long as KVA load does not exceed generator rating even at high PF.

3. As far as possible unbalanced loads for three phases should be avoided as imbalance deteriorates efficiency by cube of imbalance amount.
4. Short circuit strength when the plant is operating DG sets is lower as compared to mains distribution. In this situation, overloads and harmonics distortion have a pronounced effect on distribution within the plant.
5. Waste heat: As DGs give an electrical output of 33 - 35 %, heat losses in exhaust gases and cooling water should be utilized to improve their efficiency.
6. Some other energy efficiency measures are:
   - Steady loading
   - Dust free air
   - Proper maintenance

Cogeneration

In most of the plants, where the boilers are used for generating steam, at the end of its cycle, low pressure steam is not utilised properly and sent to condenser for recycling as feed water. High temperature heat is used for driving a gas or steam turbine and low temperature heat is utilized for heating water.

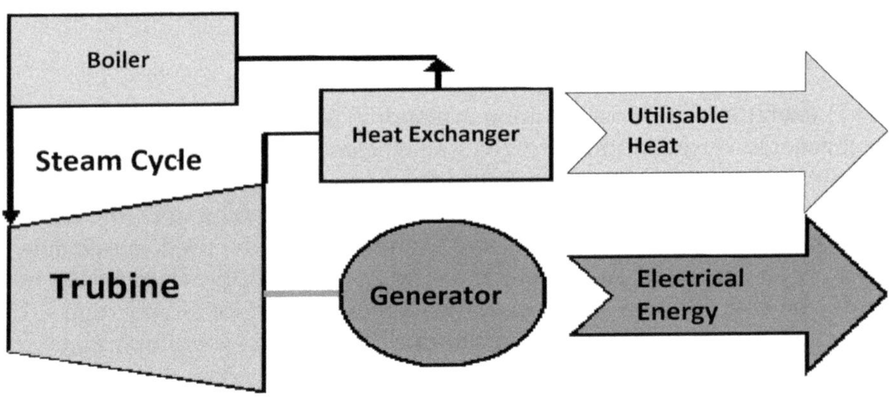

Courtesy BEE

Cogeneration is likely to be most attractive under the following circumstances:

- The demand for both steam and power is balanced
- A single plant or group of plants has sufficient demand for steam and power to permit economies of scale to be achieved

- Peaks and troughs in demand can be managed or, in the case of electricity, adequate backup supplies can be obtained from the electricity board.

Trigeneration

A subset of cogeneration is Trigeneration where cooling heating and power is obtained simultaneously.

Organic Rankine Cycle

The Organic Rankine Cycle is a thermodynamic process where heat is transferred to a fluid at a constant pressure. The fluid is vaporized and then expanded in a vapor turbine that drives a generator, producing electricity. The fluid allowsRankine cycle heat recovery from lower temperature sources such as biomass combustion, industrial waste heat, geothermal heat, solar ponds etc. The low-temperature heat is converted into useful work, that can itself be converted into electricity..

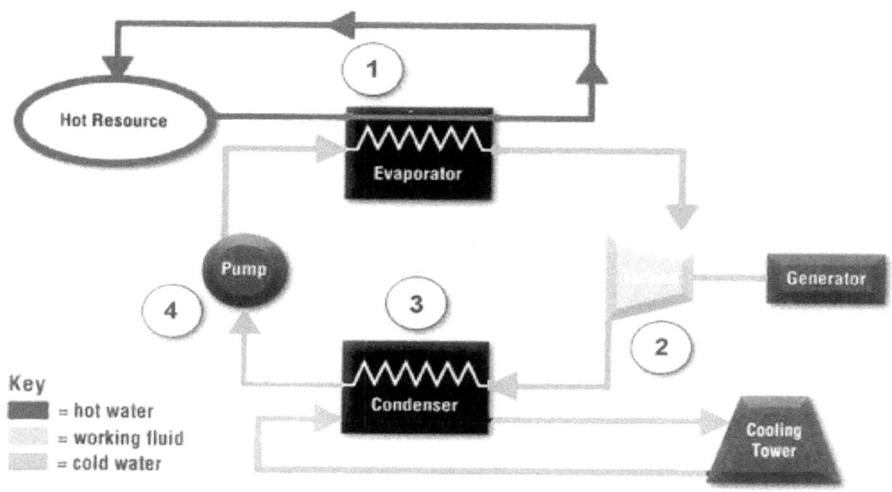

Courtesy BEE

Organic Rankine Cycle Thermadynamic Process

14

Furnaces and Refractories

Furnaces are used for melting metals and other materials for further processing. Refractories are meant for high temperature insulation. A furnace is an equipment to melt metals for casting or heat materials for **change of shape** (rolling, forging etc) **or change of properties** (heat treatment).

Due to high temperatures involved with furnaces the fuel consumption is also high and there is considerable scope for energy saving. Thus, furnaces are ideal candidates for waste heat recovery.

For good efficiency, in a given time as much as possible material should be heated to required temperature with minimum fuel.

Factors to be considered are:

- Quantity of heat to be calculated for heating of the material
- After overcoming the losses, sufficient heat to be released
- Transfer of heat from furnace gases to the stock
- Even distribution of heat to the material
- Keeping the heat losses to the minimum
- Any fuel component should not react with material

Performance of Furnace

While all efforts are made in design to ensure that heat generated from fuel combustion is passed to the material, there still are many losses as the heat is passes into walls and hearth of the furnace, loss of heat through radiation and convection from the surface apart from openings and cracks etc. Every time the furnace door is opened some heat escapes through it. Loss also takes place through free gases.

Thus, the efficiency of the furnace is determined by the amount of heat needed per unit of material.

Thermal efficiency = Heat in stock ÷ Heat from fuel consumed

If Q is quantity of heat passed to the material

$$Q = m \times Cp\ (t1 - t$$

Where Q = Quantity of heat in material in Kcal

m = Weight of material in Kg

Cp = Specific heat of material

t1 = Final temperature of material in °C

t = Initial temperature of material in °C

Indirect Method

Similar to the method adopted in boiler efficiency of furnace is measured by calculating the losses. These are:

- Heat loss in flue gases
- Loss due to evaporation of moisture in material
- Loss due to evaporation of moisture in fuel
- Heat loss due to openings
- Heat loss due to walls and surface of furnace
- Miscellaneous other losses

Fuel Economy Measures

- Ensuring complete combustion with minimum excess air
- Proper heat distribution
- Maintaining appropriate temperature
- Minimising losses due to openings
- Optimum capacity utilization
- Waste heat recovery
- Proper insulation, ceramic coating etc

15

Auxiliary Power Consumption in Large Boilers

All boilers require auxiliary equipment for their operation. Basically, these include pumps for feed water, fans for combustion air, preparation and supply of fuel among others. With boiler size increase such as power plant boilers, the auxiliaries are more and more complex.

This auxiliary equipment may consume 6 to 14 percent of energy generated from the boiler where steam is used for generating electric power. National average computed by CEA is 8.32%. Boilers for other uses to meet the requirement of steam in sugar industry, paper and pulp etc may require additional electric power to run these auxiliaries. For energy efficiency in a plant it is necessary to ensure that auxiliary equipment uses minimum power to operate optimally.

Thus, all efforts should be made to minimize consumption for auxiliaries. Even a small fraction in reduction can translate into huge monetary saving.

Measurement of auxiliary power consumption

Traditionally, auxiliary power consumption is not measured accurately missing out on the first step for energy saving. To proceed for energy efficiency measures, it is essential that energy consumption by individual auxiliary is measured. This level can be compared with the rated energy consumption or by taking this consumption as reference. Subsequent energy readings for the same auxiliary can indicate if there is any deterioration. If the bench mark established within the plant is available, it can indicate the efficiency of that auxiliary equipment.

In the absence of any periodic readings of energy consumption, it is not possible to establish the efficiency of auxiliary power. It is worthwhile to invest in energy meters for important auxiliary equipment. Ideally, there should be an online monitoring system which continuously records energy consumption as this system can provide moment to moment data of energy consumption. Typical break up is as below:

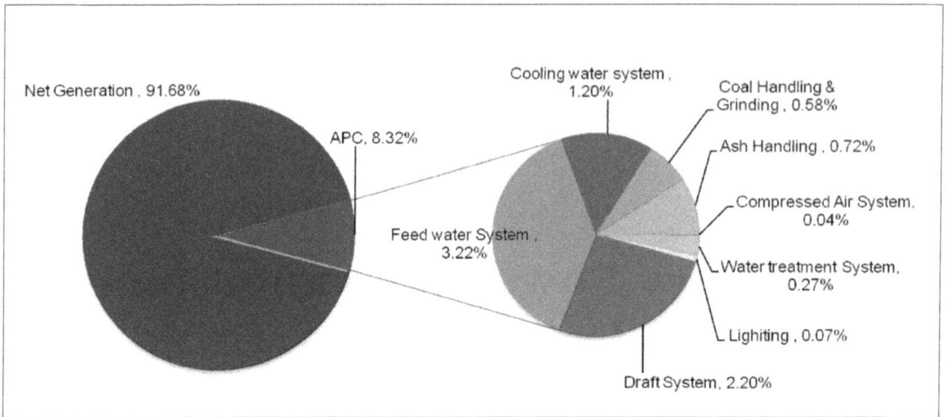

Courtesy Schneider Electric

Factors affecting APC

Following factors affect APC:

1. Plant load factor: At highest level of PLF, percentage of APC of generated power is at minimum. With low PLF, APC is a higher percentage reflecting the lowering of efficiency in generation.
2. Operational efficiency of equipment: When individual auxiliaries operate at low efficiency it results in higher APC.
3. Age of the plant: With aging, a power plant operates at lower efficiency and higher APC is observed.
4. Start up and shut down: At the time of start up and shut down APC is marginally higher till steady state of operation is achieved. Frequent starting or shutting down adds to overall APC.

When APC shows increasing trend, actions may have to be taken for individual auxiliaries:

1. Air leaks in draft system would be manifested in higher excess air. If air ingress between air pre heater and ID fans occurs.
2. Comparison of fan performance against design parameters would indicate if there are inefficiencies existing.
3. In draft system, efficiency can be improved if replacement of fans or impellers/cartridges is carried out.
4. Use of energy efficient fans, variable speed derives and high efficiency motors can minimize APC. Use of dampers and inlet guide vanes should be eliminated.

5. Analysis and reduction of Boiler Feed Water pressure and condensate extraction pressure can reduce APC. Faulty recirculation valve can also affect power consumption.
6. Coal milling and grinding system should be checked for optimized mill parameters.
7. Optimising cooling water pumping, cooling tower parameters and compressed air system will contribute to lower consumption.
8. Improvement of PF in motors operating auxiliary pumps.
9. Reducing stack temperature.

16

Waste Heat Recovery in Industries

Introduction

All the plants which use heat in production process (virtually all) have a part of heat which cannot be practically utilized. This unutilized heat component is termed as waste heat.In the past, waste heat in industries was often dumped in atmosphere. With rising energy costs, waste heat recovery in industries has in recent past become important.

Earlier, waste heat recovery as a technique was put into practice in some specific areas in boilers, furnaces, kilns etc. For instance, most of the industrial boilers have economizers for recovery of heat from flue gases exiting after combustion from the boiler and this heat is used for pre heating of boiler feed water. Similarly, furnaces use recuperators for heat recovery from exhaust gases. There are many such examples of heat recovery which have been practiced in industry for several decades. This heat recovery has been routinely carried out at high temperature usually above 350 degrees Celsius and involves large quantity of heat. Any heat recovery at lower temperatures or smaller quantities was largely ignored.

Any sub system for heat recovery has to be economical in installation as well as operation. In the past, industry did not consider medium temperature and moderate quantity waste heat as a feasible proposition for recovery. However, in the recent few years with large increases in energy prices, more and more processes would qualify as feasible for waste heat recovery. Air to air heat recovery is another concept which is now feasible.

In short, waste heat recovery has to be commercially viable.

Sources of Waste Heat and their Potential

S. no.	Source	Quality
1.	Heat in flue gases	The higher the temperature, the greater the potential value for heat recovery
2.	Heat in vapour streams	As above but when condensed, latent heat also recoverable.
3.	Convective and radiant heat lost from exterior of equipment	Low grade – if collected may be used for space heating or air preheats.
4	Heat losses in cooling water.	Low grade – useful gains if heat is exchanged with incoming fresh water.
5	Heat losses in cooling water.	a) High grade if it can be utilized to reduce demand for refrigeration. b) Low grade if refrigeration unit used as a form of heat pump..
6	Heat stored in products leaving the process	Quality depends upon temperature
7	Heat in gaseous and liquid effluents leaving process	Poor if heavily contaminated and thus requiring alloy heat exchanger.

Drivers for waste heat recovery

- Energy efficiency: When waste heat is recovered and used, the overall energy efficiency of the process gets enhanced.
- Fuel cost saving: Utilisation of waste heat in a boiler, furnace or any process results in lower fuel consumption.
- GHG emissions: As burning of fuel is reduced, GHG emissions will also reduce proportionately and lesser quantities of gases will be released.
- Enhanced equipment life.
- There is huge potential for recovering energy from air conditioning exhaust and reusing it in air conditioning load.

Quality of waste heat

As mentioned above, any sub system for waste heat recovery necessarily has to be economically viable which is determined from the quality and quantity of waste heat available from the process. Quality largely refers to the temperature at which the waste heat is available – higher the temperature the more useful would be the recovery.. The table below gives an indication of the grades of waste heat generation from various processes.

High temperature	600° to 1500° C	Metal refining furnaces, glass melting, cement kiln etc
Medium temperature	230° to 650° C	Boiler, engine and turbine exhausts, heat treatment etc
Low temperature	25° to 230° C	Cooling water outlet from air compressors, refrigeration/air conditioning, steam condensate etc

Quantity of waste heat

Quantity of waste heat available would be relevant to ensure economic feasibility. A mathematical calculation of waste heat availability is as given below:

$$\text{Quantity of heat (in kcal)} = V \times \varrho \times Cp \times \Delta t$$

Where V is flow rate of substance in m^3 / hr

ϱ is density of flue gas in kg / m^3

Cp is specific heat of substance in kcal/kg ºC

Δt is temperature difference in ºC

Typical sources of waste heat

Heat in waste gases from industrial processes (High temperature)

- Waste heat emanates from direct fuel fired processes such as metal refining or heating furnaces, cement kilns, glass melting, hydrogen plants or solid waste incinerators.
- Temperatures are typically in the range of 650 to 1500 ºC.

Combustion flue gas (Medium temperature)

- It is the most common source as majority of the industries use fuel in boilers, furnaces or kilns.
- Quantity of flue gases is substantial as it is 15 to 20 times the fuel quantity
- Temperatures are high – typically in the range of 230 to 450 ºC.
- The waste heat is readily available in the plant and, therefore, more convenient to use.
- Some constraints which have to be considered while using flue gases are as follows:
    - Flue gases usually contain gases like CO, CO2, SO2 etc which can in combination with moisture at lower temperatures form acids and corrode the metals. Heat exchangers made of special alloys may be necessary.

- Flue gases contain some contaminants which may foul the pipes and other passages.
- Heat exchangers and other heat transfer equipment tend to result in drop in pressure so it may need to be supplemented by external means.

Low temperature heat recovery

- Available from diesel generating sets, baking or curing ovens, air conditioning and refrigeration, air compressors, process steam condensate etc.
- Usually difficult to extract heat directly from the source and can be useful only as preheating arrangement.

Other sources of waste heat

- Water cooled chillers reject a significant amount of heat energy through cooling towers. This energy is at very low temperature for direct use.

Uses of waste heat

- Waste heat can be put to use depending upon the type of plant and category of heat available particularly with relation to temperature and quantity.
- Pre heating of combustion air: For improving the efficiency of combustion preheated air can be obtained by using waste heat. This is the most common use to which waste heat is utilized. This method can result in reducing energy consumption in boilers and other direct fuel fired equipment.
- Pre heating of boiler feed water: This is very commonly done in most of the boilers because pre heating of feed water will reduce the fuel consumption resulting in enhanced energy efficiency. In the case of closed loop steam systems condensate water is already at sufficiently high temperature; there, preheating would be relevant for makeup water.
- Vapour Absorption Refrigeration: In the facilities or industries which require refrigeration or air conditioning of high volume and simultaneously have waste heat available, vapour absorption systems are a preferred option.
- Pre heating for process requirements: many chemical and other industrial require hot fluids, air or water. Waste heat can be most appropriately used.

Benefits of waste heat recovery

- Direct benefits: Recovery of waste heat results in reduced energy consumption and consequent increase in energy efficiency.

- Indirect benefits:
  - Reduction in pollution as toxic combustible waste burnt in incinerators are used for heat recovery
  - Reduction in equipment size as flue gas produced are in reduced quantities and consequently result in reduction in sizes of fans, stacks, ducts etc.
  - Reduction in auxiliary energy consumption as reduced size of equipment results in reduction in energy consumption.

Development of waste heat recovery system

- Compatibility of waste heat quality: Quality of waste heat energy must be better than or equal to energy required for end use. High temperature usage would require waste heat of matching temperature. However, low or medium temperature use can be met with high temperature heat
- Scheduling: Quite often, generation of waste heat may not be continuous. Similarly usage may also be intermittent. It will then be necessary to match requirement with availability where waste heat availability or usage is intermittent. Maximum efficiency of waste heat utilization would be achieved when generation and usage are matched. For example, if boiler flue gases are used for heating its combustion air or feed water, the arrangement would be ideal because both are part of the same system and come into use whenthe boiler is operating. If the schedule of waste heat generation and utilisation does not match, some form of thermal storage would be necessary. Thermal storage is expensive and not a preferred option.
- Location: Physical proximity between availability of waste heat and usage is ideal for achieving best efficiency. Long distances for transferring heat in any form would result in expensive system installation.

Trends in waste heat recovery devices

Conventional devices: As waste heat recovery has been practiced in industries for many years (even before the energy crisis), many conventional devices for waste heat recovery are in use. For the sake of completeness, these have been described along with newer systems.

Recuperators: Heat exchange takes place between flue gases and air through metallic or ceramic walls. Ducts or tubes carry air which has to be preheated. The other side contains waste heat in the form of flue gases. Stack losses are decreased by lower gas temperatures and also by discharging reduced quantities of exhaust gases due to fuel savings. Heat transfer from hot gases to the inner surface takes place through radiation while heat transfer to cold air takes place through convection.

Alternately, hot gases are passed through a number of parallel tubes of small diameter whereas incoming air passes through outer shell surrounding the tubes. Incoming air passes over hot tubes a number of times in direction normal to the axis of tubes.

Regeneration: Heat transfer by regeneration is ideal when large capacities are involved as in glass and steel melting furnaces. Size of the regenerator depends upon time between reversals, thickness and conductivity of brick, and heat storage ratio of bricks. Long periods between reversals would mean higher thermal storage and hence higher cost.

Economisers: In case of boilers, economizer is used to utilize waste heat from flue gases to pre heat boiler feed water. A thumb rule is that for every 22ºC (64ºF) reduction in flue gas temperature by passing through an economizer, there is 1% saving in fuel in boiler.

Waste heat boilers: These are ordinary water tube boilers in which hot exhaust gases from gas turbines, incinerators etc pass over a number of parallel tubes containing water. Water gets evaporated in the tubes and collects in steam drum from which it is drawn off for use in heating or process steam. In gas based power plants, waste heat boilers are used after gas turbine generation to further operate a steam turbine to generate power. Thus, the efficiency of a gas based power plant is enhanced by almost 50%.

Heat pumps: In this device, after a fluid flows through an evaporator it is completely vapourised. Low temperature vapour is compressed with external energy input raising its temperature and pressure. The vapour flows through a condenser giving up its energy to become a liquid. This liquid is expanded through an expansion valve back to the evaporator. Heat pumps have the ability to upgrade heat to a value twice that of energy consumed.

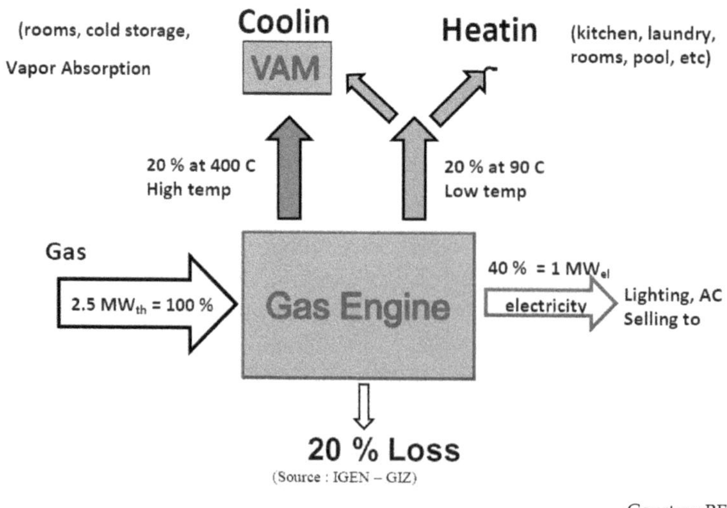

Courtesy BEE

Heat pumps are useful where both heating and cooling are simultaneously required. One such example is plastic industry where water after cooling injection moulding machine is used to provide space heating.

Heat wheels: These are widely used in low to medium temperature heat recovery systems particularly in HVAC. A disc rotates between hot air duct and cold air duct which are installed side by side. As the disc rotates heat is transferred from hot air to cold air.

Heat wheels are manufactured and widely exported from India. Their use in our country is also becoming popular.

Trigeneration: A relatively new concept which gives rise to simultaneous production of electricity, heat and cold Trigeneration involves simultaneous production of electricity, heat and cold. Main components of a trigeneration system are alternators driven by gas engine, VAM for cooling and warm water for heating. Trigeneration is estimated to have efficiency higher than cogeneration by 30 to 50%. Thus, trigeneration has a total efficiency around 80%. The system can be useful in small to large scale with typical range of 250kW to several MWs. This system is ideal for buildings and industries which use simultaneous electricity, heating and cooling.

Trigeneration can ideally be used with dedicated or independent power plant and group or district CCHP. In India, FDI is freely allowed and there is no need for industrial licence. Eight US and European countries have already entered Indian market.

New trends in R&D (Source : Dr Satish Kumar)

- Thermoelectrics:

  Heat converted directly into electricity

- Low temperature organic rankine cycle:

  Organic, high molecular mass fluid with a liquid-vapor phase change, or boiling point, occurring at a lower temperature than the water-steam phase change

- Thermally activated air conditioning such as absorption and adsorption chillers.

- Kalina cycle: It uses typically two fluids

  Thermodynamic process for converting thermal energy from the industrial waste heat stream into re-useable mechanical power.

Conclusion: Waste heat recovery has become an economic necessity in industries as it can result in fuel or energy use reduction. Simultaneously, it can reduce the carbon footprint of the industry by reduced emissions. While the conventional waste heat recovery devices have a continuing validity, newer devices are being developed to help in further development of heat recovery and energy efficiency.

17

Renewables and Environment

In the present times, environmental issues are being highly emphasised which would require new industrial plants as well as existing units to adhere to environmental regulations. At the same time, renewables do offer economy in obtaining energy from these sources. This has happened due to technical developments in this field and increasing dependence on renewables world wide which has brought economy of scale as well. A major scheme of energy efficiency from Bureau of Energy Efficiency namely Perform, Achieve and Trade (PAT) provides huge incentive to designated (energy intensive) industries by keeping aside energy from renewables while calculating energy consumption post retrofits for energy efficiency. Thus, industries could profitably adopt renewable energy and keep within environmental parameters also.

Industrial plants typically are highly conducive for solar energy generation due to huge areas available on roof tops which also result in insulation and reduced ingress of heat from roofs. Land requirement for solar is also avoided when roof tops are used for solar energy. Plants requiring hot water for their use would find solar energy inexpensive for the purpose. Similarly, refrigeration requirement could also be met from solar energy.

To a lesser extent than solar, it is possible to use wind energy. Wind energy is not limited by daylight hours but location of wind turbines may be a limiting factor. Thereafter, convenient wheeling in for power would also be required

A few precepts for using solar energy

- Thumb rule for generating energy from solar is that for 1KW power, a panel of 1 square meter is required. Conversely, from roof area available for installing solar panel, electrical output can be calculated.
- Ideally, solar energy may be used as available in daylight. If it is required after daylight hours storage devices will be required.

- For requirement of hot water, solar heating can be used directly. For obtaining AC voltage, an inverter has to be provided.
- Simple payback calculation of roof top solar installation largely depends on the cost of PV cells and to a small extent on the cost of structure. It is assumed here that there is no requirement for investing in land.

18

Energy Efficiency in Industrial Buildings

Basic principles of energy efficiency for buildings apply to industrial buildings as well. While Energy Conservation Building Code (ECBC) has been designed for commercial buildings, it could be referred to for industrial buildings as well even though such buildings do not fall in that category.

Industrial buildings may have conditioned or non-conditioned spaces. Conditioned buildings in India are mainly air conditioned and for energy efficiency purpose the focus would be on air conditioning as it consumes most of the energy required. On the other hand, non conditioned buildings would require measures which help in minimizing ingress of heat from outside and control of internal heat sources.

Most often, reasons for energy wastage in buildings are due to neglect and lack of effort by the organization to take some basic measures which will be described in subsequent paragraphs. Certain measures should be implemented when the buildings are designed whereas others are possible in existing buildings.

EE Measures for Existing Buildings

Some of the measures are listed below:

1. Measurement of consumption: This is a basic requirement for effective energy efficiency to have an accurate data for consumption on a regular basis. Monitoring of the data provides indications of wastages and inefficient use of energy. For this purpose each building should have metering arrangement.
2. Use of renewable - solar roof top: Solar energy from panels fitted on roof of the building can help in reducing consumption from conventional sources.

3. If conventional lights are fitted these should be replaced with energy efficienct ones such as LEDs, sodium vapour among others. It is observed that the industrial buildings often have irregular or non-uniform lux levels. Excessive lux level would result in wastage. An audit of lighting should be undertaken to ensure that uniform lighting level is available.
4. Ways to avoid ingress of external heat: Minimising ingress through glass by use of double glass, insulating roof top/ using white paint, controlling air ingress through doors and other openings.

EE Measures for buildings being designed or planned

Measures are listed below:

1. At design or planning stage, it is advisable to ensure that building envelope is best suited for energy efficiency. For Indian conditions, effect of heating by sun should be minimized by choosing the most ideal orientation for the building – solar heat gain coefficient (SHGC)
2. Check of fenestration and openings should be made for best energy efficiency – U Factor.

Air Conditioning

In any building, a large amount of energy – 50 to 60 %- is consumed by AC plant. AC calls for focused attention for energy efficiency. For details, chapter on air conditioning should be referred to.

Lighting

Next to air conditioning, maximum energy consumption is normally attributable to Lighting. The Chapter on lighting energy efficiency may be referred to.

19

Energy Monitoring, Energy Balance and Practising ISO 50001

For energy efficiency in a plant the basic requirement is to know the energy input and energy consumed in the plant. The difference between these two figures will provide the energy lost or wasted. The basic precept is that it is not possible to control unless the basic data is available. In most of the industries, energy consumption figures when related to production, will provide specific energy consumption (SEC). SEC can be used to monitor increase or decrease periodically and for comparison with other plants

While it is easy to ascertain the energy supplied to a plant, as energy meter is provided by the supplier, periodically read and bill raised. It is more difficult to know the consumption in the plant and consequently the loss or wastage.

Calculation of energy consumption

For calculating energy consumption, it is necessary to provide energy meters for all the feeders from the main supply so that consumption in all the branches is available. When the consumption in all feeders is added together, it will provide the energy consumption for the whole plant.

With stand alone meters for feeders, it is not possible to obtain accurate data of energy loss as energy level in plants keeps changing. To overcome this problem, it is necessary to have on line monitoring system which can provide accurate data in real time enabling loss figures to be available.

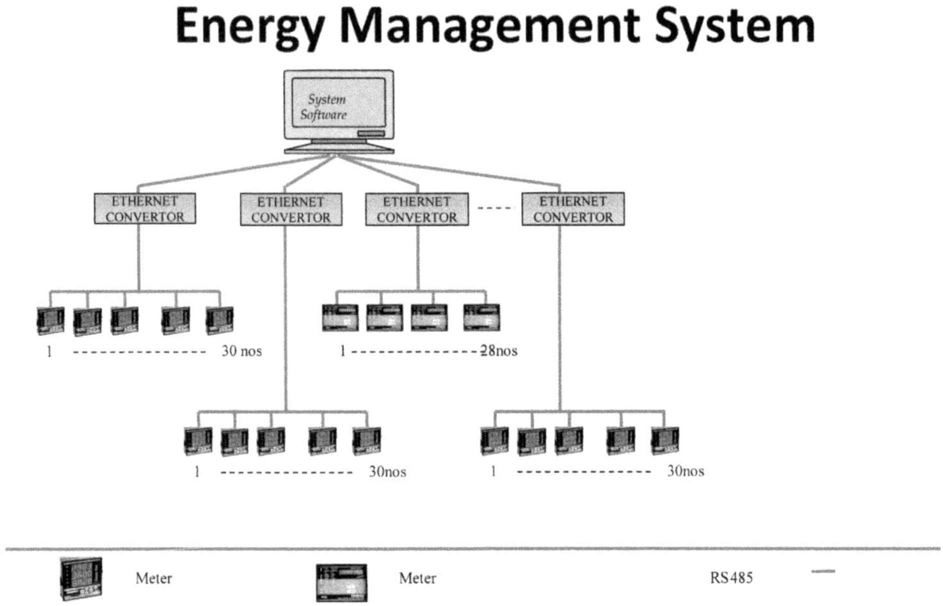

Courtesy Conzerv

The system shown above can also provide for mechanical data such as temperature. In ISO 50001, EnMS is included which enables an organization to monitor energy consumption although it is not necessary to have an online system. Benefits accruing from online system are obvious and if the cost involved is not too high online system providing data in real time should be preferred.

Energy Balance

For the purpose of energy accounting, efficiency calculation and SEC, the first step is energy balance. If 100 units of energy are supplied to a plant, it should be possible to determine where these 100 units have been consumed and the units which have been lost. In a large plant, SEC should be calculated for each step of production or process. Consequently, energy balance calculation will also have to be undertaken for each step. In this effort overall consumption will not have much use. Metering for each process and online monitoring will be the right solution.

Applications of Energy Management System

- Benchmarking: With monitoring system in place, it is possible to determine accurately the energy consumed per unit of production. Continuous monitoring can provide SEC and from this data it is possible to arrive at minimum energy consumption per unit which can be the benchmark for production or process for comparison. Any excesses can be analysed against this benchmark.

- Predictive Maintenance: If any equipment draws excess energy for same output, it may indicate a maintenance problem to be addressed. For example if an electric motor is drawing excess current there could be problem with winding, bearing or the load. If that is allowed to continue the motor can get burnt.
- Root Cause Analysis: If a breakdown does occur then system can indicate the cause of breakdown.
- Power Quality: Multi function meters can provide indication of power quality mainly PF and harmonic content. If any load or circuit, shows low PF or high level of harmonics with this system, it is possible to attend to and correct the problem.
- Measurement and Verification: As M&V is for accurately determining savings achieved when an energy saving measure is implemented with same conditions before and after, the system makes it easy to do M&V.
- Cost Allocation: Quite often the companies have to do internal cost allocation for a process or production. The system can provide data to enable the exercise.
- Smart Power Management: Internal power management is possible through the system. For example, any process which is not in series can be performed when the tariff is low or excess power is available.
- Energy Resource Planning: Whenever it is planned to enhance or reduce the supply of energy say through electricity board, the usual practice is to come to the figure by ad hoc means or estimates. The system enables the plant to more accurately plan for increasing or decreasing the demand.
- Energy Balancing: Accurate energy balancing is possible with the system.
- Breaker Status – remote monitoring: It is possible through the system by simple on off sensor.
- Automated Energy Billing System for Townships: With appropriate metering the system can be use for energy billing in a multistory building, township or a spread out residential facility.

Advantages of Energy Monitoring

1. Meters provide raw data only and it is not possible to correlate the readings to production or proper utilization (in non production organizations). On line monitoring and loss detection is only possible with such a system.
2. The system can also indicate power quality. If multimeters are provided, power factor and even harmonics level can also be available for appropriate action.

ISO 50001 for Energy Efficiency

It is a specification created by the International Organization for Standardization (ISO) for an energy management system. The standard specifies the requirements for establishing, implementing, maintaining and improving an energy management system, whose purpose is to enable an organization to follow a systematic approach in achieving continual improvement of energy performance, including energy efficiency, energy security, energy use and consumption. The standard aims to help organizations continually reduce their energy use, and therefore their energy costs and their greenhouse gas emissions.

ISO 50001 was released in June 2011 and is suitable for any organization – whatever its size, sector or geographical location.[1]

A significant feature in ISO 50001 is the requirement to "...improve the EnMS and the resulting energy performance". ISO 50001, therefore, has made a major leap forward in 'raising the bar' by requiring an organization to demonstrate that they have improved their energy performance. There are no quantitative targets specified - an organization chooses its own then creates an action plan to reach the targets. With this structured approach, an organization is more likely to see some tangible financial benefits.

The main objective of the standard is to improve energy-related performance and energy efficiency continuously and to identify energy reduction opportunities. This systematic approach will help organizations to establish systems and processes.

Consistent energy management helps organizations to realize untapped energy efficiency potential. They will benefit from cost savings and make a significant contribution to environmental and climate protection, for example by the permanent reduction of CO_2 emissions. The organization can discover potential savings and competitive advantages.

ISO 50001 provides a framework of requirements that help organizations to:

- develop a policy for more efficient use of energy
- fix targets and objectives to meet the policy
- use data to better understand and make decisions concerning energy use and consumption
- measure the results
- review the effectiveness of the policy and
- continually improve energy management.

ISO50001 certification proves that the energy management system meets the requirements of the stsndard. This gives customers, stakeholders, employees and management more confidence that the organization is saving energy. It also helps to ensure that the energy management system is working throughout the organization.

20

Initiation of Energy Efficiency in Plant Utilities

In this chapter, certain basic steps are outlined for initiating energy efficiency in plant utilities.

Energy Audit

Energy audit is most well known way in the last two decades for carrying out an assessment of energy consumption in a plant, factory, building or any other facility which consumes energy..

Types of energy audits

- The commonest way is energy audit of the entire plant which covers all energy consuming or load points existing in the plant. This kind of audit is most desirable when no prior information or data is available for energy consumption in the plant. It is also relevant as a policy measure for periodic assessment of energy consumption in a plant.
- Energy audit may also be carried out for specific equipment within a plant such as motor audit, audit for compressed air or steam. This usually takes place at a subsequent period after an overall energy audit has been carried out and it is established that a focus is necessary for certain equipment such as motors, compressed air etc as gaps are observed in energy efficiency in these areas.
- There are also specialised audits other than energy audits such as Harmonics analysis.

Conduct of energy audits

- Energy audits are conducted in many varied ways which more often than not depend on the service providing organization..Energy auditors may conduct the audit and at end of their work provide a report which analyses all sub systems in a plant indicating the scope of energy savings in these areas. These auditors may be qualified individuals, small or medium firms or very large organizations. Quite often some of these carry out implementation of their recommendations and arrive at the targeted energy consumption or efficiency thus validating the assured savings.

- Energy audits are also conducted by large manufacturers of equipment when replacement of equipment or sub system is envisaged. For example, old refrigeration system could be replaced by state of the art equipment providing certain assured energy savings. It could also be tied to payment through energy savings which is an incentive for the client as he may not have to make any upfront payment for the replacement.
- Sometimes the energy audit is conducted to fulfill specific mandated condition of any authority or financing organisation. In such situations qualified energy auditors may be engaged for cost saving to obtain the requisite certification.

Audit Report

Energy audit report will consist of the following:

- The methodology adopted for the audit including the tests and results thereof.
- Areas and equipment where energy losses are evident and quantum thereof.
- Possible energy savings and means of achieving such savings. The means could be improved operation, repair/renovation or replacement.. These can be applicable to systems, sub systems or any specific equipment.
- Audit report must include the cost of achieving energy savings as well as savings.
- The vendors or manufacturers who can provide the requisite services or replacement should be mentioned in audit report.

Preparation for Energy Audit

- Data logging which is carried out during the operation prior to the audit should be made available for audit as that can help the auditors to determine the level at which an equipment was operated as well as the energy consumption.
- Any variations in energy supply such as voltage levels etc in the recorded data would also be useful to the auditors.
- As far as possible the plant should be operated at production level which is normally obtaining in the plant. Any extremes of low level operation or abnormally high production level can be misleading..

Action on outcome of Audit

When the audit report is received, the management should constitute a group to carry out the implementation. Basic approach should be to obtain maximum energy savings with minimum savings so that payback is fastest. Depending on the financial health of the company, plan should be formulated to carry out the measures for renovation, repair or replacement. Some relevant factors to be considered are as follows:

- If the terms of reference for audit include implementation, then the work could be assigned to them. However, competitive bids may be obtained from other vendors who can carry out these jobs.
- If the audit firm is also a manufacturer then replacement can be assigned to them and they can be committed for showing the savings in the report.
- In case the above conditions are not included for auditors prior to audit, they could still be involved to measure the actual savings for a fee once the work of repair renovation or replacement is carried out.
- Energy audit provides important data and information about the behavior of equipment in actual operation. The readings obtained during audit could be used as benchmark for the energy consumption and other relevant measurements. Any deterioration after the audit can indicate action required for bringing the equipment to appropriate level in operation.

www.ingramcontent.com/pod-product-compliance
Ingram Content Group UK Ltd.
Pitfield, Milton Keynes, MK11 3LW, UK
UKHW020047200726
13845UKWH00044B/28

9 789389 569193